T0171943

SpringerBriefs in Physics

SpringerBriefs in Physics are a series of slim high-quality publications encompassing the entire spectrum of physics. Manuscripts for SpringerBriefs in Physics will be evaluated by Springer and by members of the Editorial Board. Proposals and other communication should be sent to your Publishing Editors at Springer.

Featuring compact volumes of 50 to 125 pages (approximately 20,000–45,000 words), Briefs are shorter than a conventional book but longer than a journal article. Thus, Briefs serve as timely, concise tools for students, researchers, and professionals.

Typical texts for publication might include:

- A snapshot review of the current state of a hot or emerging field
- A concise introduction to core concepts that students must understand in order to make independent contributions
- An extended research report giving more details and discussion than is possible in a conventional journal article
- A manual describing underlying principles and best practices for an experimental technique
- An essay exploring new ideas within physics, related philosophical issues, or broader topics such as science and society

Briefs allow authors to present their ideas and readers to absorb them with minimal time investment.

Briefs will be published as part of Springer's eBook collection, with millions of users worldwide. In addition, they will be available, just like other books, for individual print and electronic purchase.

Briefs are characterized by fast, global electronic dissemination, straightforward publishing agreements, easy-to-use manuscript preparation and formatting guidelines, and expedited production schedules. We aim for publication 8–12 weeks after acceptance.

More information about this series at http://www.springer.com/series/8902

Eugen Mircea Anitas

Small-Angle Scattering (Neutrons, X-Rays, Light) from Complex Systems

Fractal and Multifractal Models
for Interpretation of Experimental Data

 Springer

Eugen Mircea Anitas ⓘ
Bogoliubov Laboratory
of Theoretical Physics
Joint Institute for Nuclear Research
Dubna, Moscow, Russia

ISSN 2191-5423   ISSN 2191-5431   (electronic)
SpringerBriefs in Physics
ISBN 978-3-030-26611-0   ISBN 978-3-030-26612-7   (eBook)
https://doi.org/10.1007/978-3-030-26612-7

This Springer imprint is published by the registered company Springer Nature Switzerland AG
The registered company address is: Gewerbestrasse 11, 6330 Cham, Switzerland

Preface

This book addresses the basic physical phenomenon of small-angle scattering (SAS) of neutrons, X-rays or light from complex hierarchical nano- and microstructures. The emphasis is on developing theoretical models for the material structure containing self-similarity or fractal clusters. Within the suggested framework, key approaches for extracting structural information from experimental scattering data are investigated and presented in detail. The text includes the selection of fractal models, ranging from mass to surface fractal type, along with their combinations. The range of parameters which can be obtained paves the road toward a better understanding of the correlations between geometrical and various physical properties (electrical, magnetic, mechanical, optical, dynamical, transport, etc.) in fractal nano- and micro-materials. This has important consequences in the design and fabrication of complex materials with predefined structure and functions. Thus, the book is aimed at a broad range of graduate students and researchers in physics, chemistry, biology and materials science.

Since the underlying property of the investigated systems is self-similarity, fractal geometry is used to provide the proper language for the description of structural properties derived from SAS analysis. In this framework, an intrinsic pattern may repeat either exactly or statistically under a transformation of scale. The former case gives rise to exact self-similar (deterministic) fractals, where the shape of the smaller parts resembles exactly the shape of the overall object, while the later one generates structures where only the statistical properties remain invariant (random/stochastic fractals).

Although most of the fractal structures and phenomena found in nature have a random character, recent advances in materials science and nanotechnology have given scientists and engineers a strong impetus on creating various deterministic fractals at nano- and microscales. Such structures are tailored for very specific tasks, and their tunable geometrical properties play an essential role in improving their performances (mainly mechanical and transport ones) up to a few orders of magnitude. Without losing from generality, we focus here on building deterministic fractal models and investigating their structural properties since in a good approximation they can be used to study the properties of random fractals as well.

This approach allows us an analytic description of geometrical properties determined from SAS data, such as the radius of gyration or the number of objects forming a fractal, and thus it provides us with a very powerful theoretical tool in structural characterization of the new generation of deterministic nanomaterials.

Due to the fast-growing literature and space constraints, here are presented only the essential developments instead of a full literature review. Therefore, the book is organized as follows.

Chapter 1 gives a short overview of SAS from fractals and provides a gentle exposure to the main theoretical methods and models used for experimental data analysis. The presented results can be applied to neutron, X-ray as well as light scattering. The advantages and limitations in extracting structural properties are outlined.

Chapter 2 introduces a few important issues concerning definition, classification and properties of fractals and multi-fractals. Several theoretical approaches to generate such structures are described, and the main methods to characterize them are introduced, with a focus on dimension spectra.

Chapter 3 presents in detail the general principles of SAS technique and illustrates them on scattering from simple Euclidean objects. We describe the connection between the structure of an object and its corresponding scattering intensity. The effects of polydispersity and of orientational averaging are presented and discussed.

Chapter 4 illustrates the modeling approach used for structural characterization of fractals and explains how fractal properties can be derived from SAS intensity. We introduce several prototypical fractal models which can be used to explain the main types of fractal power-law decays observed experimentally: mass, surface, fat and multi-fractals.

Chapter 5 briefly summarizes the models and the structural properties which can be obtained from a SAS experiment. Also, some perspectives for future research directions and limitations in this field are outlined.

The author is indebted to many colleagues for valuable discussions and particularly to Dr. A. Yu. Cherny, Prof. V. A. Osipov and Dr. A. I. Kuklin for their valuable discussions

Dubna, Russia   Eugen Mircea Anitas
April 2018

Contents

Chapter 1
Introduction

Abstract A short overview of the most recent results on SAS from deterministic fractals is presented and the fundamental distinction between SAS from mass and surface fractals is emphasized. The behavior and key properties of classical scattering models such as simple power-law or Beaucage are explained, indicating their most common applications and limitations. The important role played by deterministic fractal models in extracting additional structural information from SAS data is underlined.

1.1 Overview

One of the fundamental challenge in materials science and nanotechnology is understanding the correlations between structural and physical properties of materials such as mechanical [1], optical [2, 3], electrical [4], magnetic [5], statistical [6], transport [7] or dynamical [8] ones. This knowledge enables the design and fabrication of advanced materials with predefined structures and functions for a wide range of applications spanning electrochemical release of cancer cells [9], cardiac electrotherapy [10], fractional transport of photons in deterministic aperiodic structures [11], smart tissues [12, 13] intracellular production of hydrogels and synthetic RNA granules [14], drug delivery in pharmaceuticals [15], polymer coatings [16], estimation of effective density of engineered nanomaterials for *in vitro* dosimetry [17] or production of magnetorheological suspensions and elastomers [18–22] for various applications. Therefore, a detailed characterization of the spatial distribution of inhomogeneities represents an important step in manufacturing of advanced materials with improved physical properties and functions.

Although there are a number of different methods for the structural analysis of nano/micro-meter particles, basically all of them can be classified either as imaging (real-space analysis) or diffraction/scattering (reciprocal-space analysis). Imaging techniques such as electron microscopy (EM) provide images with high resolution but the method has the inherent disadvantage that special sample preparation is required, since samples need to be in a vacuum environment. In addition, technical limitations of EM such as the interference with external magnetic fields or the need

© The Author(s), under exclusive license to Springer Nature Switzerland AG 2019
E. M. Anitas, *Small-Angle Scattering (Neutrons, X-Rays, Light)*
from Complex Systems, SpringerBriefs in Physics,
https://doi.org/10.1007/978-3-030-26612-7_1

for a highly stable power source may significantly affect the results and make their maintenance a labor-intensive process.

On the other hand, scattering techniques although being severely limited by the loss of phase information, have several distinct advantages which makes them very useful in investigating nanoscale morphologies: samples usually don't need any additional preparation, the obtained physical quantities are averaged over a macroscopic volume, structures as large as proteins and viruses can be analyzed, and basic structural properties can be obtained without any model or approximation. In particular for fractal systems [23], i.e. structures which look similar under a transformation of scale, small-angle scattering (SAS) of neutrons (SANS), X-rays (SAXS) or light (SALS) [24] is very effective since it can distinguish between mass and surface fractals through the value of the scattering exponent in the fractal regime (see Sec. 3.1).

Theoretically, a convenient framework which can describe the scaling behavior of such systems is provided by fractal geometry [23], which makes use of two main classes of fractal systems: exact self-similar/deterministic fractals, where each part replicates exactly the structure of the whole, and statistical/random self-similar fractals, where only statistical properties replicates at various scales of magnification.

Deterministic fractals are generally built either by a top-down (TD) approach, where according to an initial prescribed rule, an initially arbitrarily structure is repeatedly divided into a set of smaller components of the same shape, or by a bottom-top (BT) approach, where the process is reversed. One of the most common example of TD is that of a one-dimensional Cantor set, where one considers first a line segment which is divided into three equal parts, the middle third is removed, and the process is repeated *ad infinitum* for each remaining segments. Following this process, at every iteration one obtains a structure self-similar to the previous iteration, which is identical upon rescaling. Due to their exact self-similar property, most (if not all) of the deterministic fractals are generated artificially, such as Menger sponge-like [25], molecular Sierpinski hexagonal gasket [26], two-dimensional Cantor sets [27], centrally/peripherally/uniformly distributed circular fractals [28] or three-dimensional octahedral structures [29].

Random fractals on another hand, are naturally occurring and have been observed in various physical systems ranging from galactic clusters down to molecular and atomic structures such as polymers and interfaces, or even to elementary particles. However, in a good approximation we can study the structure of random fractals by using a deterministic fractal with the same fractal dimension. This is a very common approach since the early times of fractals and has been used successfully in many studies. For example, it has been shown that the transfer across random fractal surfaces is very close to the response of deterministic model geometries [30] or that modification of the construction algorithm of a three-dimensional exact self-similar Koch snowflake, significantly improves the agreement between calculated and experimental SANS spectra for fractal supernucleosomal chromatin structures [31]. Moreover, by introducing polydispersity in deterministic fractals one obtains the same SAS intensity as from random fractals [32–37].

In addition, this approximation has the advantage that is more convenient in numerical calculations and, probably what is the most important, it allows an analytical description of various properties and of geometrical parameters such as the scattering intensity spectrum or radius of gyration [33]. Also, recent advances in nanotechnology have allowed researchers to prepare various artificially nanoscale deterministic fractal objects [26, 27, 29] and therefore for this type of systems this approximation becomes exact. Therefore, using models of deterministic fractal systems provides us not only with an increasing important tool in describing random fractals but also an exact framework that gives us 'exactly solvable models' for describing the properties of the new generation of exact self-similar nano/micro materials.

Indeed, very early the regularity of deterministic fractals, using the SAS method, was the subject of several papers [2, 38]. However, the important question of which information could be extracted from such scattering intensities was only partially addressed, and basically the structural information was limited to the fractal dimension and, in certain conditions, to the outer and inner size of the fractals. The full potential of using deterministic fractals was not clearly understood until recently, when it has been shown in Ref. [33] that by using the deterministic fractal approach one can obtain a lot more of structural information (besides of those obtained using random fractal approach) from SAS experiments, such as the scaling factor, the number of fractal iterations and the total number of structural units from which the fractal is composed.

In the same reference, it has been obtained also an analytic expression for the scattering intensity and for fractal radius of gyration, therefore greatly simplifying the description and understanding of fractals. More recently, the 'deterministic approach' to the analysis of scattering intensity $I(q)$, where q is the scattering wavevector, has been extended to surface fractals as well, and it was shown that the range of parameters which can be extracted is the same, although the nature of the scaling factor is different [39, 40]. For both mass and surface fractals, the scaling factor is obtained from the log-periodicity of the curves $I(q)q^{D_m}$ and respectively $I(q)q^{2d-D_s}$ as a function of q on a double logarithmic scale, where d is the Euclidean dimension of the space in which the fractals are embedded D_m is the mass fractal dimension, and D_s is the surface fractal dimension (see next section). While for mass fractals the log-periodicity arises from the self-similarity of *distances* between the scattering objects, for surface fractals the log-periodicity arises from the self-similarity of *sizes* of scattering objects [39].

1.2 Motivation and Research Objectives

Most physical systems such as porous solids, macromolecular solutions or colloidal suspensions consist of a complex mix of hierarchical morphologies, all of which having a contribution to the scattering intensity on all length scales from about 1 nm up to few microns. In general, whenever a material has inhomogeneities larger that 1 nm the corresponding scattering patterns at small angles become observable. The patterns are characterized by an asymptotic exponential region at small values of q, where

$I(q) \propto q^0$, and by regions with a power-law decay of scattering intensity at high q, where $I(q) \propto q^{-\tau}$. The scattering exponent τ is related to the Euclidean dimensionality of the object: $\tau = 4$ for three-dimensional objects (e.g. spheres, cubes), $\tau = 2$ for two-dimensional objects (e.g. thin disks), and $\tau = 1$ for one-dimensional objects (e.g. thin rods).

However, in many cases the exponent τ takes non-integer values between one and four, which have been related to power-law distribution of scattering units [41], and therefore the structure of such materials is analyzed by using the concept of fractal geometry. For such systems, the scattering intensity in the fractal region can be written as [42]:

$$I(q) \propto q^{D_s-2(D_m+D_p)+2d}. \tag{1.1}$$

For a mass fractal $D_s = D_m < d$ and $D_p = d$, and Eq. (1.1) becomes:

$$I(q) \propto q^{-D_m}. \tag{1.2}$$

For surface fractals $D_m = D_p = d$, $d - 1 < D_s < d$, and D_p is the dimension of the "pores" surrounding the mass fractal. Therefore, Eq. (1.1) can be represented as:

$$I(q) \propto q^{-(2d-D_s)}. \tag{1.3}$$

As noted before, the true power of SAS technique resides in its ability to provide both dimensions D_m and D_s experimentally.

A large class of SAS experimental data may show not only a single power-law decay, but a *succession* of power-law decays (typically two or three) with various scattering exponents, such as precipitated silica [43], single-walled carbon nanotubes suspensions [44] or composite elastomeric membranes [45]. For such types of systems, a simple power-law model as given by Eqs. (1.1), (1.2) or (1.3) is not enough to reveal also the sizes of the basic objects forming the fractal or of a given level of organization inside the fractal, such as atoms, molecules or aggregates. To describe such hierarchically organized systems, one usually resort to the Beaucage model [46], which can describe both Guinier and Porod regions, and can be used to obtain the radii of gyration and scattering exponents for each structural level. In general for an arbitrarily number n of interrelated structural levels, it reads as:

$$I(q) \simeq \sum_{i=1}^{n} \left(G_i e^{-q^2 R_{g,i}^2/3} + B_i e^{-q^2 R_{g,i+1}^2/3} \left(\frac{erf\left(qk R_{g,i}/\sqrt{6}\right)}{q} \right)^{\tau_i} \right), \tag{1.4}$$

where i is the structural level of the largest size, R_g is the radius of gyration at level i, $G_i \equiv c^2 N_p I_e$ are the exponential prefactors, B_i are constant prefactors specific to the type of power-law scattering as determined by by the regime in which τ_i falls, erf is the error function, and k is an empirical constant and accounts for an approximation involved in the description of the power-law at low q. For steep power-law decays, i.e.

when $\tau > 3$, $k = 1$ while for weak power-law decays $k = 1.06$. For x-rays scattering N_p is the number of particles in the scattering volume, I_e is the scattering factor for a single electron, and c is the number of electrons inside the particle. Also, for a Porod law $B = 2\pi N_p c^2 S_p I_e / V_p^2$, where V_p and S_p is the volume and respectively the surface area of the particle, while for a Gaussian polymer ($\tau = 2$), $B = 2G_i / R_{g,i}$.

More recently, it has been shown in Ref. [47] that the practice of allowing both Guinier and Porod prefactors to vary independently in a nonlinear least-squares fits give rise to undesired artifacts in the Beaucage model. As a result, a new empirical model, known as Guinier-Porod model has been introduced to overcome such inconveniences [48]. However, it still does not reproduce peaks in experimental data and it can not handle oscillations characteristics to monodisperse deterministic fractals.

A solution to this problem has been proposed by Cherny et al. in Ref. [33] where deterministic mass fractal models have been suggested for analysis of SAS data characterized by a generalized power-law decay, i.e. a superposition of maxima and minima on a simple power-law decay. Such models have the advantage that they enable explicit analytic solutions for the scattering amplitude and are characterized by the absence of phenomenological parameters, which are often employed in modeling random fractals. They allow a number of additional parameters to be extracted from the scattering intensity, such as: the scaling factor, iteration number or the total number of objects forming the fractal. For more general class of structures, containing structures of mass fractals whose dimension depends on the iteration number, a model based on deterministic *fat* fractals has been suggested in Ref. [34] for explaining a *succession* of generalized mass fractals power-law decays. It is known that the simple power-law decay (or a succession of them) can be recovered by introducing the polydispersity in the sizes of fractals. This makes the idealized models of deterministic fractals even more useful for modeling random fractals since almost any physical sample shows a certain degree of polydispersity.

Here, we present several general methods to generate deterministic models, and perform a detailed analysis of their properties as obtained from SAS curves. We underline their main features which can be exploited when analyzing experimental SAS data. The approach presented allows for various types of generalization including different shapes and sizes of fractals and/or of their component objects, thus giving new insight into the structure of a large class of complex hierarchical materials.

References

1. Fan, J.A., Yeo, W.H., Su, Y., Hattori, Y., Lee, W., Jung, S.Y., Zhang, Y., Liu, Z., Cheng, H., Falgout, L., Bajema, M., Coleman, T., Gregoire, D., Larsen, R.J., Huang, Y., Rogers, J.A.: Nat. Commun. **5**, 3266 (2014)
2. Allain, C., Cloitre, M.: Phys. Rev. B **33**(5), 3566 (1986)
3. Sotelo, J.A., Pustovit, V.N., Niklasson, G.A.: Phys. Rev. B **65**(24), 245113 (2002)
4. Boukheir, S., Len, A., Füzi, J., Kenderesi, V., Achour, M.E., Éber, N., Costa, L.C., Oueriagli, A., Outzourhit, A.: J. Appl. Pol. Sci. **134**, 8 (2017)

5. Szczerba, W., Costo, R., Veintemillas-Verdaguer, S., Morales, M.d.P., Thünemann, A.F.: IUCr. J. Appl. Cryst. **50**(2), 481 (2017)
6. Luscombe, J.H., Desai, R.C.: Phys. Rev. B **32**(3), 1614 (1985)
7. Balankin, A.S., Golmankhaneh, A.K., Patiño-Ortiz, J., Patiño-Ortiz, M.: Phys. Lett. A **382**(23), 1534 (2018)
8. Galiceanu, M.: J. Phys. A: Math. Theor. **43**(30), 305002 (2010)
9. Zhang, P., Chen, L., Xu, T., Liu, H., Liu, X., Meng, J., Yang, G., Jiang, L., Wang, S.: Adv. Mater. **25**(26), 3566 (2013)
10. Xu, L., Gutbrod, S.R., Ma, Y., Petrossians, A., Liu, Y., Webb, R.C., Fan, J.A., Yang, Z., Xu, R., Whalen, J.J., Weiland, J.D., Huang, Y., Efimov, I.R., Rogers, J.A.: Adv. Mater. **27**(10), 1731 (2015)
11. Dal Negro, L., Inampudi, S.: Sci. Rep. **7**(1), 2259 (2017)
12. Bica, I., Anitas, E.M.: Composites Part B **159**, 13 (2019)
13. Bica, I., Anitas, E.M.: Mat. Sci. Eng. B **236-237**, 125 (2018)
14. Nakamura, H., Lee, A.A., Afshar, A.S., Watanabe, S., Rho, E., Razavi, S., Suarez, A., Lin, Y.C., Tanigawa, M., Huang, B., DeRose, R., Bobb, D., Hong, W., Gabelli, S.B., Goutsias, J., Inoue, T.: Nat. Mater. **17**(1), 79 (2017)
15. Alford, A., Kozlovskaya, V., Kharlampieva, E.: Biological Small Angle Scattering: Techniques, Strategies and Tips, Barnali, C., Ines, G.M., Shuo, Q., Volker, S.U. (eds.) 1st edn. (Springer, Singapore, 2017), Chap. 15, pp. 239–258
16. Hughes, A.E., Trinchi, A., Chen, F.F., Yang, Y.S., Cole, I.S., Sellaiyan, S., Carr, J., Lee, P.D., Thompson, G.E., Xiao, T.Q.: Adv. Mater. **26**, 4504 (2014)
17. Deloid, G., Cohen, J.M., Darrah, T., Derk, R., Rojanasakul, L., Pyrgiotakis, G., Wohlleben, W., Demokritou, P.: Nat. Commun. **5**, 3514 (2014)
18. Bica, I., Anitas, E.M., Chirigiu, L., Daniela, C., Chirigiu, L.M.: Colloid. Polym. Sci. **296**, 1373 (2018)
19. Bica, I., Anitas, E.M.: J. Ind. Eng. Chem. **64**, 276 (2018)
20. Bica, I., Anitas, E.M.: Mater. Des. **155**, 317 (2018)
21. Bica, I., Anitas, E.M., Lu, Q., Choi, H.J.: Smart Mater. Struct. **27**(9), 095021 (2018)
22. Gavrilovici, A.M., Anitas, E.M., Chirigiu, L., Bica, I., Negrutiu, M.L.: Adv. Pol. Tech. (2019)
23. Mandelbrot, B.B.: The Fractal Geometry of Nature. Freeman, W.H (1982)
24. Feigin, L.A., Svergun, D.I.: Structure Analysis by Small-Angle X-Ray and Neutron Scattering. Springer, Boston (1987). https://doi.org/10.1007/978-1-4757-6624-0
25. Mayama, H., Tsujii, K.: J. Chem. Phys. **125**(12), 124706 (2006)
26. Newkome, G.R., Wang, P., Moorefield, C.N., Cho, T.J., Mohapatra, P.P., Li, S., Hwang, S.H., Lukoyanova, O., Echegoyen, L., Palagallo, J.A., Iancu, V., Hla, S.W.: Science (New York) **312**(5781), 1782 (2006)
27. Cerofolini, G.F., Narducci, D., Amato, P., Romano, E.: Nanoscale Res. Lett. **3**(10), 381 (2008)
28. Şandru, A.: J. Mod. Optics **59**(3), 199 (2012)
29. Berenschot, E.J.W., Jansen, H.V., Tas, N.R.: J. Micromec. Microeng. **23**, 055024 (2013)
30. Filoche, M., Sapoval, B.: Phys. Rev. Lett. **84**(25), 5776 (2000)
31. Ilatovskiy, A.V., Lebedev, D.V., Filatov, M.V., Petukhov, M.G., Isaev-Ivanov, V.V.: J. Phys.: Conf. Ser. **351**(1), 012007 (2012)
32. Schmidt, P.W.: J. Appl. Cryst. **24**(5), 414 (1991)
33. Cherny, A.Y., Anitas, E.M., Osipov, V.A., Kuklin, A.I.: Phys. Rev. E **84**(3), 036203 (2011)
34. Anitas, E.M.: Eur. Phys. J. B **87**, 139 (2014)
35. Anitas, E.M.: Adv. Cond. Mat. Phys. **2015**, 501281 (2015)
36. Anitas, E.M., Slyamov, A., Todoran, R., Szakacs, Z.: Nanoscale Res. Lett. **12**(1), 389 (2017)
37. Anitas, E.M., Slyamov, A.M.: Ann. Phys. **530**(11), 1800187 (2018)
38. Schmidt, P.W., Dacai, X.: Phys. Rev. A **33**(1), 560 (1986)
39. Cherny, A.Y., Anitas, E.M., Osipov, V.A., Kuklin, A.I.: IUCr. J. Appl. Cryst. **50**(3), 919 (2017)
40. Cherny, A.Y., Anitas, E.M., Osipov, V.A., Kuklin, A.I.: Phys. Chem. Chem. Phys. **19**(3), 2261 (2017)
41. Schmidt, P.W.: J. Appl. Crystallogr. **15**(5), 567 (1982)

42. Pfeifer, P., Ehrburger-Dolle, F., Rieker, T.P., González, M.T., Hoffman, W.P., Molina-Sabio, M., Rodríguez-Reinoso, F., Schmidt, P.W., Voss, D.J.: Phys. Rev. Lett. (2002)
43. Schaefer, D.W., Justice, R.S.: Macromolecules **40**(24), 8501 (2007)
44. Schaefer, D.W., Brown, J.M., Anderson, D.P., Zhao, J., Chokalingam, K., Tomlin, D., Ilavsky, J.: J. Appl. Cryst. **36**(3), 553 (2003)
45. Anitas, E.M., Bica, I., Erhan, R.V., Bunoiu, M., Kuklin, A.I.: Rom. J. Phys. **60**(5–6), 653 (2015)
46. Beaucage, G.: J. Appl. Cryst. **28**(6), 717 (1995)
47. Hammouda, B.: J. Appl. Cryst. **43**(4), 716 (2010)
48. Hammouda, B.: J. Appl. Cryst. **43**(6), 1474 (2010)

Chapter 2
Fractals: Definitions and Generation Methods

Abstract This chapter recalls in Sect. 2.1 the basic notions from fractal theory which are necessarily for understanding the concepts introduced throughout the book. The main mathematical concepts and definitions of fractals and multifractals that arise from set theory are considered in Sects. 2.2 and 2.3, with an emphasis on the interpretation methodology. Several general methods for generating a rich variety of fractal structures are reviewed in Sect. 2.4, and some examples will be used in Chap. 4 to calculate the small-angle scattering intensity.

2.1 Introduction

2.1.1 Set Theory

We consider a general $n - dimensional$ Euclidean space $\mathbb{R}^n$ with elements denoted by lower case letters x, y, etc. In coordinate form, the points are written as $x = (x_1, x_2, \ldots, x_n)$, $y = (y_1, y_2, \ldots, y_n)$. The used metric on $\mathbb{R}^n$ is the distance [1]

$$|x - y| = \left(\sum_{i=1}^{n} |x_i - y_i|^2 \right)^{1/2}, \tag{2.1}$$

where x_i and y_i, with $i = 1, \ldots, n$ are the coordinates of x, and respectively of y. Recall that a metric is a function that defines a distance between each pair of elements of a set. Formally, a *metric* on a set S is a function $d : S \times S \rightarrow [0, \infty)$, where $[0, \infty)$ is the set of non-negative real numbers and for all $x, y, z \in S$ the following conditions are satisfied [1]:

1. $d(x, y) \geq 0$, non-negativity or separation axiom,
2. $d(x, y) = 0 \Leftrightarrow x = y$, identity of indiscernibles,
3. $d(x, y) = d(y, x)$, symmetry,
4. $d(x, z) \leq d(x, y) + d(y, z)$, sub-additivity or triangle inequality,

© The Author(s), under exclusive license to Springer Nature Switzerland AG 2019   9
E. M. Anitas, *Small-Angle Scattering (Neutrons, X-Rays, Light)*
from Complex Systems, SpringerBriefs in Physics,
https://doi.org/10.1007/978-3-030-26612-7_2

that is, the distance between distinct points is positive, the distance from x to y is the same as the distance from y to x, and the distance from x to z via y is at least as great as from x to z directly. A metric on a space induces topological properties like open and closed sets. A set with a metric is called a metric space. In Euclidean geometry the shortest distance between two points is a line. When the usual metric given by Eq. (2.1) is replaced by a new metric, other types of geometries are obtained.

Thus, the *closed* and *open balls* centered at point x and with radius r are defined by $B(x, r) = \{y : |y - x| \le r\}$, and respectively $B^0(x, r) = \{y : |y - x| < r\}$. The set of points within the distance δ of a set S is called the $\delta - neighborhood$ of S, and is expressed by $S_\delta = \{x : |x - y| \le \delta$ and $y \in S\}$.

If the elements of an infinite set S can be written as $x_1, x_2, \ldots$ where each element of S appears at a certain place in the list, then the set S is said to be *countable* (e.g. $\mathbb{Z}$ and $\mathbb{Q}$) otherwise S is *uncountable* (e.g. $\mathbb{R}$). The *supremum* sup S of a non-empty set of real numbers is the least number m for which every element $x \in S$, satisfies $x \le m$. If no such element exists in S, then the supremum is infinity. In an analogous manner, the *infimum* inf S is the greatest number p such that for every $x \in S$, $p \le x$. Note that neither the supremum nor the infimum need not to belong to the set S. The *diameter* $|S|$ of the set S is defined as the greatest distance between the elements of S, and is written as $|S| = \sup\{|x - y| : x, y \in S\}$. If $|S|$ is finite, the set S is *bounded*. If for every element $x \in S$, $\exists B(x, r)$ of positive radius that is contained within S, then S is *open*, otherwise is *closed*. The set S is called a neighborhood of point x if there $\exists B(x, r) \subset S$. The smallest closed set containing S is known as the *closure* of S, denoted by $\bar{S}$, while the largest open set contained in S is called the *interior* of S and is denoted int S [1]. The concepts of openness and closedness generalize the idea of an open interval in the real line. In practice, they provide a fundamental way to speak about nearness of points in a topological space, that is, to define other properties such as continuity, connectedness or compactness.

Several examples are given below to illustrate the above definitions: a non-empty set S is closed but not open, with $\bar{S} = S$, and int $S = \emptyset$. The interval (a, b) with a and $b \in \mathbb{R}^1$ and $a < b$ is open but not closed, with $\overline{(a, b)} = [0, 1]$ and int $(a, b) = (a, b)$. The interval $[a, b]$ is closed but not open, with $\overline{[a, b]} = [a, b]$ and int $[a, b] = (a, b)$. The half-open interval $[a, b)$ is neither open nor closed, with $\overline{[a, b)} = [a, b]$ and int $[a, b) = (a, b)$. The set $S = \{0, 1/1, 1/2, 1/3, \ldots\}$ is closed but not open, with $\bar{S} = S$ and int $S = \emptyset$.

The *boundary* ∂S of the set S is defined by $\partial S = \bar{S} \setminus$ int S, and therefore an element $x \in \partial S$ if and only if the closed ball $B(x, r)$ intersects the set S and its complement $\forall r > 0$. The set P is *dense* in S if there exists elements in P arbitrarily close to every point in S. i.e. $S \subset \bar{P}$. The set S is *compact* if is both closed and bounded, i. e. if an arbitrarily collection of open sets that covers the set S contains a finite sub-collection which also covers S. A set S is *connected* if it consists of only one single "part", i.e. there do not exists any open sets A and B such that $A \cup B$ contains S, where $S \cap A$ and $S \cap B$ are non-empty and disjoint sets. The *connected component* of x is the largest connected subset of S which contains the element x. When the connected component of each point in S consists of just that point, the set S is *totally disconnected* [1].

For example, the middle third Cantor set is compact and totally disconnected. Indeed, it can be written as $C = \bigcap_{k=0}^{\infty} G_k$, where G_k is the union of 2^k disjoint closed intervals in $[0, 1]$, each with length 3^{-k}. Here, k represents the iteration number. At each step k, the union G_k is closed since is the union of finitely many closed sets. Taking into account that the intersection of any collection of closed sets is closed, it turns out that C is also a closed set, since C is a subset of the interval $[0, 1]$. Since C is also bounded, with $|C| = 1$, then C is a compact set. Since C is closed, we have $\bar{C} = C$, and since C contains no open interval, int $C = \emptyset$, and therefore $\partial C \equiv \bar{C} \setminus int C = C$, that is, the boundary of the middle Cantor set is the set itself. Furthermore, if we consider two different elements $x, y \in C$ with $x < y$, it is possible to find an G_k such that $x \in [a, b]$ and $y \in [c, d]$ with $a < b < c < d$. If we consider p such that $b < p < c$, and since C consists from the union of disjoint open intervals such as $(-1, p)$ and $(p, 2)$ with $x \in (-1, p)$ and $y \in (p, 2)$, it follows that the middle third Cantor set is totally disconnected.

2.1.2 Measures Theory

In order to develop a systematic way to assign a number to each subset of a set, one has to use the concept of a measure of a set, such as Lebesgue measure on the Euclidean space, and which assigns the conventional length, area and volume to subsets of the n-dimensional Euclidean space $\mathbb{R}^n$. Therefore, intuitively this is interpreted as its size. This assignment shall be built in such a way that when a set S is decomposed into a finite number of parts, then the sum of the sizes of the parts equals the size of the whole.

Technically, μ is a measure on $\mathbb{R}^n$ if it satisfies the following properties [1]:

1. $\mu(\emptyset) = 0$, i.e. the empty set has zero measure,
2. $\mu(S) \leq \mu(P)$ if $S \subset P$, i.e. the larger the set, the larger the measure,
3. $\mu\left(\bigcup_{i=1}^{\infty} S_i\right) \leq \sum_{i=1}^{\infty} \mu(S_i)$, if $S_1, S_2, \ldots$ is a finite sequence of sets, i.e. the sum of the measures of the parts is at least equal to the measure of the whole. The equality holds when all S_i are disjoint Borel sets.

Let's consider an element $a \in \mathbb{R}^n$ together with a unit point mass concentrated at a, and show that the quantity defined by $\mu(S) \equiv 1$ if $a \in S$ and 0 otherwise, is a measure. From the definition $\mu(\emptyset) = 0$ if $a \notin \emptyset$. Then, if we suppose that $a \in A$ and $b \in B$ with $A \subset B$, $\mu(A) = \mu(B) = 1$. Also, if $a \notin \emptyset$, $\mu(A) = 0 \leq \mu(B)$. Thus, in both cases the condition $\mu(A) \leq \mu(B)$ if $A \subset B$ is satisfied. Finally, if we consider that $S_1, S_2, \ldots$ is a sequence of sets, then when $a \notin S_i, \forall i \in \mathbb{N}$, we have $a \notin \bigcup_{i=1}^{\infty} S_i$ and thus $\mu(\bigcup_{i=1}^{\infty} S_i) = 0 = \sum_{i=1}^{\infty} \mu(S_i)$. Conversely, when $a \in S_j$, $j \in \mathbb{N}$, then $a \in \bigcup_{i=1}^{\infty} S_i$ so that $\mu(\bigcup_{i=1}^{\infty} S_i) = 1 = \mu(S_j) \leq \sum_{i=1}^{\infty} \mu(S_i)$. If S_i are disjoint, then $a \notin S_i$ for $i \neq j$ so that $\mu(S_i) = 0$ for $i \neq j$ and thus $\mu(\bigcup_{i=1}^{\infty} S_i) = 1 = \sum_{i=1}^{\infty} \mu(S_i)$. This shows that all the above conditions defining a measure are satisfied and hence μ is a measure.

The *support* of μ, denoted spt μ is the smallest closed set Y such that $\mu (\mathbb{R}^n \setminus Y) = 0$, i.e. the support of a measure is a set on which the measure is concentrated. An element x belongs to the support if and only if $\mu (B(x, r)) > 0$, $\forall r > 0$. Thus, if the set S contains the support of μ, then μ is a measure on S.

The concept of a measure can be used to model mass distributions, since this approach allows point/line/surface masses as well as masses described through volume density functions. Formally, a measure on a bounded subset of $\mathbb{R}^n$ for which $0 < \mu(\mathbb{R}^n) < \infty$ is called a *mass distribution*, and we will refer to $\mu(S)$ as the mass of the set S. As an example, we consider the Lebesgue measure on $\mathbb{R}^n$ and define the set $S = \{(x_1, x_2), \ldots, x_n) \in \mathbb{R} : a_i \leq x_i \leq b_i\}$ as the *coordinate parallelepiped* in $\mathbb{R}^n$. The corresponding n-dimensional volume will be given by $\text{vol}^n (S) = (b_1 - a_1)(b_2 - a_2) \ldots (b_n - a_n)$, and the n-dimensional Lebesgue measure will take the form [1]:

$$\mathscr{L}^n (S) = \inf \left\{ \sum_{i=1}^{\infty} \text{vol}^n (S_i) : S \subset \bigcup_{i=1}^{\infty} S_i \right\}, \tag{2.2}$$

that is, the Lebesgue measure is obtained by considering all coverings of S by countable collections of coordinate parallelepipeds, and take the smallest total volume possible. When $n = 1$, the coordinate parallelepiped becomes a simple interval, then $\text{vol}^1 \equiv L$ represents the usual length. Thus, the 1-dimensional Lebesgue measure can be written as:

$$\mathscr{L}^1 (S) = \inf \left\{ \sum_{i=1}^{\infty} (b - a) : S \subset \bigcup_{i=1}^{\infty} [a_i, b_i] \right\}, \tag{2.3}$$

where $S = \bigcup_i [a_i, b_i]$ is a finite union of disjoint intervals. By considering that the line segment of length L has an associated measure μ such that $\mu(A) = \mathscr{L}^1(L \cap A)$, where $A \subset \mathbb{R}^2$, then μ is a mass distribution with support L since $\mu(A) = 0$ for $A \cap L = \emptyset$, and is evenly spread along the line segment of length L. The same unit mass can be split equally between the two intervals of L_1, with $L_1 \subset L$, or it can be split equally between the two subintervals in L_2, with $L_2 \subset L_1$. By repeating the same procedure for higher steps k we can construct a mass distribution μ on the middle third Cantor set by repeated subdivisions, and which is distributed in the most uniformly possible way. In the reminder of the book, we shall deal with Hausdorff measures, which are generalizations of Lebesgue measures to dimensions s that have no integral values.

2.1.3 Hausdorff Measure

A countable collection of sets $\{S_i\}$ of diameter at most δ that cover a subset M of $\mathbb{R}^n$, that is, $F \subset \bigcup_{i=1}^{n} S_i$ and $0 < |S_i| \leq \delta$ for each i is called a δ-cover of M. If $s > 0$, then for each $\delta > 0$ we can define [1]:

$$\mathcal{H}^s_\delta (M) = \inf \left\{ \sum_{i=1}^{\infty} |S_i|^s : \{S_i\} \text{ is a } \delta-\text{cover of M} \right\}, \tag{2.4}$$

that is, we look at all covers of M by sets of diameter at most δ and minimize the sum of sth powers of the diameters. As we decrease the value of δ, the range of possible covers of M is also decreased, and as a consequence $\mathcal{H}^s_\delta (M)$ increases, approaching a limit. For any subset M of $\mathbb{R}^n$ this limit exists and is called the s-*dimensional Hausdorff measure* of M, and formally [1],

$$\mathcal{H}^s (M) = \lim_{\delta \to 0} \mathcal{H}^s_\delta (M). \tag{2.5}$$

In defining the s-dimensional Hausdorff measure, the sets in the covering are considered arbitrarily, but they may be taken either open or closed. However, they will give the same measure, although the approximations $\mathcal{H}^s_\delta (M)$ may be different.

When $s \in \mathbb{N}$, then Eq. (2.5) gives the n-dimensional Hausdorff measure, which coincide up to a constant with the n-dimensional Lebesgue measure given by Eq. (2.2). Thus, if M is a Borel subset of $\mathbb{R}^n$, then [1]

$$\mathcal{H}^n (M) = c_n^{-1} \text{vol}^n (M), \tag{2.6}$$

where $C_n = \pi^{n/2} / 2^n (n/2)!$ if n is even, and $c_n = \pi^{(n-1)/2} ((n-1)/2)! / n!$ if n is odd. Here, the coefficients c_n represent the volume of an n-dimensional ball of diameter 1. Therefore, $\mathcal{H}^0 (M)$ represents the number of points in M, $\mathcal{H}^1 (M)$ is the length of a smooth curve in M, $\mathcal{H}^2 (M) = (4/\pi) \times area(M)$ if M is a smooth surface, etc.

One of the most important properties of Hausdorff measures concerns its scaling, and it states that if $f : \mathbb{R}^n \to \mathbb{R}^n$ is a similarity transformation of scale factor $\lambda > 0$, and $M \subset \mathbb{R}^n$, then

$$\mathcal{H}^s (f(M)) = \lambda^s \mathcal{H}^s (M). \tag{2.7}$$

Hausdorff measures are also translation and rotation invariant, that is $\mathcal{H}^s (M + y) = \mathcal{H}^s (M)$, where $M + y = \{x + y : x \in M\}$. This holds since the function f is a congruence, i.e. if $|f(x) - f(y)| = |x - y|, \forall x, y \in \mathbb{R}^n$ then $\mathcal{H}^s (f(M)) = \mathcal{H}^s (M)$.

2.2 Fractal Dimension

Most of the times we are interested in how much space a set occupies near to each of its points. For example, curves (e.g. Weierstrass function) and boundaries (e.g. coastlines, cluster's surface) can behave so irregularly, that is they have a pathological behaviour, that any intrinsic meaning associated to their well-defined length or tangents is lost, since their values depend on the method of measurement and thus they increase without bound. To overcome this issue, we generally measure a set such that irregularities are recorded at various scales δ, and then we investigate the behavior of these measurements as the scale tends to zero. This approach leads to the notion of *fractal dimension*, which provides a statistical index of complexity. There exists several formal mathematical definitions of fractal dimensions. From a mathematical point of view, the Hausdorff dimension [2] is probably the most important since it is defined for any set and it is introduced by a precise definition (although quite abstract) based on measures that are convenient to work with. However, for analyzing experimental data and for mathematical computations we have to resort to a more convenient approach. The most common one is to use the box-counting dimensions, which are based on the concept of change in detail with a change of scale [3].

Generally, both approaches lead to the same value of the fractal dimension s, especially when s is a critical boundary between growth rates that are insufficient to cover the space, and growth rates that are overabundant. Therefore, for fractals that occur in nature or are created artificially by modern nanotechnologies, and which are the main structures investigated in this book, the Hausdorff and box-counting dimensions coincide.

2.2.1 Formal Definition: Hausdorff Dimensions

From Eq. (2.4) we can see that for a given set $M \subset \mathbb{R}^n$, $\mathscr{H}^s(M)$ is a non-increasing function of s when $\delta < 1$, and therefore $\mathscr{H}^s(M)$ (see Eq. (2.5)) is also non-increasing. Moreover, if we consider that $t > s$ and $\{S_i\}$ is a δ-cover of M, then we can write that $\sum_i |S_i|^t \le \sum_i |S_i|^{t-s} |S_i|^s \le \delta^{t-s} \sum_i |S_i| s$, and therefore considering infimums over all δ-covers, we obtain that $\mathscr{H}_\delta^t(M) \le \delta^{t-s} \mathscr{H}_\delta^s(M)$. Setting $\delta \to 0$ it is clear that if $\mathscr{H}^s(M) < \infty$, then $\mathscr{H}^s(M) = 0$ for $t > s$. This shows that it exists a critical value at which $\mathscr{H}^s(M)$ changes from ∞ to 0. This value is called *Hausdorff dimension* [2], and formally it can be written as:

$$\dim_H M = \inf \left\{ s \ge 0 : \mathscr{H}^s(M) = 0 \right\} = \sup \left\{ s : \mathscr{H}^s(M) = \infty \right\}, \qquad (2.8)$$

so that we have:

$$\mathscr{H}^s(M) = \begin{cases} \infty, & \text{if } 0 \le s < \dim_H S \\ 0, & \text{if } s > \dim_H S. \end{cases} \qquad (2.9)$$

One of the basic properties of the Hausdorff dimension is that for any given $s > 0$, there are uncountable fractals with Hausdorff dimension s in the n-dimensional Euclidean space $\mathbb{R}^n$, $(n \ge \lceil s \rceil)$. This is known as the *Hausdorff Dimension Theorem* [1] and it expresses the fact that to a given value of the fractal dimension there correspond an infinite number of fractals. Another important property is that if $S \subset \mathbb{R}^n$, then the range of values for the fractal dimensions is $0 \le \dim_H S \le n$. This condition arise immediately from the properties of Hausdorff measures.

We shall present in the following, an example of calculating the Hausdorff dimension of a simple fractal set. Usually, this process implies calculation of an upper and a lower estimate of the fractal dimension, which involve first a geometric calculation. Let's consider the 1-dimensional middle Cantor set C and try to show using rigorous arguments that its fractal dimension is $s = \log 2 / \log 3$. We denote the intervals which form the sets G_k in generating C, *level-k* intervals. Then, G_k consists of 2^k level-k intervals, each one having the length 3^{-k}. For the δ-cover of C we consider the intervals of G_k, and therefore $\mathscr{H}^s_{3^{-k}}(C) \leq 2^k 3^{-ks}$. The right side of the inequality equals 1 when $s = \log 2 / \log 3$. By taking the limit $k \to \infty$, then $\mathscr{H}^s(F) \leq 1$. For each interval S_i we have $3^{-(k+1)} \leq |S_i| \leq 3^{-k}$, with $k \in \mathbb{N}$ and S_i a cover of C. As a consequence, S_i can intersect at most one level-k interval. This is true, since the disjunction of level-k intervals is at least 3^{-k}. For $j \geq k$, the previous inequality shows that the intervals S_i intersect at most $2^{j-k} = 2^j 3^{-sk} \leq 2^j 3^s |S_i|^s$ level-j intervals of G_j. By choosing the values of j large enough so that the condition $3^{-(j+1)} \leq |S_i|, \forall S_i$ is satisfied, then counting the intervals gives $2^j \leq \sum_i 2^j 3^s |S_i|^s$, since the covers $\{S_i\}$ intersect all 2^j intervals of length 3^{-j}. The last inequality thus becomes $\sum_i |S_i|^s \geq 1/2 = 3^{-s}$. Taking the natural logarithms of the both sides in the above equation, we can solve for s, that is $s = \log 2 / \log 3$. It is clear from this example that determination of both Hausdorff measures and dimensions may not be an easy task, even for simple fractals such as the middle Cantor set, since it requires complicated manipulations that provides little insight. This is due to the lower estimate of fractal dimension, since we have to take into account all possible δ-covers to obtain the infimum.

However, for self-similar fractals, this lengthy process can be avoided by providing technical conditions on the sequence of contractions ψ_i. Recall that a set of function ψ_i from M to itself are *contractions* on a metric space (M, d) if there exists $r_i \in \mathbb{R}$, $0 \leq r_i < 1$, such that $\forall x, y \in M$, we have

$$d(f(x), f(y)) \leq r_i d(x, y), \tag{2.10}$$

where r_i are called *contractivity factors*. Such a condition shall be sufficient for an explicit determination of the fractal dimension. The most common one is the *open set condition*, where the condition $\bigcup_{i=1}^{m} \psi_i(V) \subseteq V$ is satisfied. Here, V is a relatively compact open set and the sets in union on the left are pairwise disjoint. The open set condition ensures that images of $\psi_i(V)$ do not overlap too much. When ψ_i is also a *similitude*, that is, ψ_i is a composition of isometry and a dilation around a point, then the unique fixed point of a given ψ is a set whose Hausdorff dimension is s, where s is the unique solution of [4]:

$$\sum_{i=1}^{m} r_i^s = 1. \tag{2.11}$$

The contraction coefficient of a similitude represents the magnitude of the dilation.

We can use this result to compute again the Hausdorff dimension of the middle Cantor set, since it is self-similar and it satisfies the open set condition. Let's consider now two points a_1 and a_2 on the line $\mathbb{R}$ and let ψ_i be the dilations of ratio $1/3$ around a_i. The unique non-empty fixed point of the corresponding mapping ψ is a Cantor set and the dimension s is the solution of $(1/3)^s + (1/3)^s = 2\,(1/3)^s = 1$, which gives $s = \log 2/\log 3$. Similar arguments can be used for any self-similar set. For example, in the case of Sierpinski gasket in the plane, Eq. (2.11) becomes $(1/2)^s + (1/2)^s + (1/2)^s = 3\,(1/2)^s = 1$, which gives $s = \log 3/\log 2$, as expected.

2.2.2 Practical Definition: Box-Counting Dimensions

Naturally occurring or artificially created fractal structures usually display a power-law type scaling over a finite range of scales. By covering the fractal with disjoint boxes of varying sizes we can estimate the box-counting dimension by counting the number of boxes that contain the fractal. This method is well suited especially for planar structures and image analysis since the later ones can be processed as pixel patterns.

Let's consider first a nonempty and bounded subset $S \subset \mathbb{R}^2$ and let $N(\delta)$ denote the smallest number of sets of diameter at most δ that can cover S. Then, the box-counting dimension of the set describes the way in which the number $N(\delta)$ grows as $\delta \to 0$. If one finds that $N(\delta)$ obeys, at least approximately, a power-law decay, then one can write:

$$N(\delta) \simeq u\delta^{-s}, \qquad (2.12)$$

with $u > 0$. Solving for dimension s, gives $s \simeq \log N(\delta)/\log(1/\delta) + \log u/\log \delta$. The box-counting dimension is defined by neglecting the second term, and thus:

$$s = \lim_{\delta \to 0} \frac{\log N(\delta)}{\log(1/\delta)}, \qquad (2.13)$$

which is widely used empirically. In particular, for image analysis, determination of the fractal dimensions is performed by using the following steps:

1. All pixels belonging to the investigated object are counted.
2. The image is then divided into a mesh of squares with sizes $l_1 \times l_1 = 2 \times 2$ pixels, and the number $N(l_1)$ of squares containing pixels belonging to the structure are counted. The process is repeated for squares of size $l_2 \times l_2 = 3 \times 3$ pixels, $l_3 \times l_3 = 4 \times 4$ pixels etc., and the numbers $N(l_2)$ and $N(l_3)$ etc. are recorded.
3. The points $(l_i, N(l_i))$, $i = 1, 2, \ldots, P/2$, with P being half the size of a square image (in pixels) are plotted on a double logarithmic scale.
4. Using the least-square method, a regression line $y = a + bx$ is obtained.
5. The fractal dimension is then determined as the slope of this line b, $(b < 0)$, and thus $s = -b$. This last step can be justified by taking natural logarithms in Eq. (2.13), and thus rewriting it as $\log N(\delta) = \log u - s \log \delta$.

Some implementations of this algorithm can be found in Refs. [5, 6]. Note that an image consists always from a finite number of pixels, and any natural/artificial fractal displays self-similarity only over a limited range of scales. Also, slightly less values for the fractal dimensions are to be expected as compared with theoretical ones (where available), since the dimension of a mathematical fractal containing an infinite number of points is larger. However, the box-counting algorithm nevertheless can provide, through Eq. (2.13), a fractal dimension that is meaningful across the scales. Generally, the accuracy of the box-counting method can be improved by first filtering the image, and only then applying the previous algorithm.

The maximum value for the fractal dimension is 2 for two-dimensional structures. In the case of a plane-filling curve, its length is a constant independent of the length scale δ. For a regular shape, such as a circle of radius r, the perimeter is constant when the length scale is close to r, and thus $s = 1$. However, the curves can be so convoluted that they may completely fulfill the available space. Thus $s = 2$, and its length is linearly related to δ.

2.2.3 Equivalent Fractal Dimensions

There exists additional various ways to define box dimensions and their equivalents, with one variant being used over another, depending on the application sought. Although the box-counting dimension given by Eq. (2.13) is very convenient for computer implementations, it does not consider the frequency with which the investigated structure might be found within the covering mesh. As a consequence, the properties related to the neighborhoods of individual structures are not distinguishable. This is of particular importance in materials science, where the structure often consists from disconnected parts distributed across the investigated plane/volume. For example, in the distribution of various molecular clusters such as bys-terpyridine or dendrimer macromolecules in solution [7, 8], various discrete parts can have distinct degrees of clustering. A solution to this problem is provided by introducing the cluster dimension D_c, which takes into account the corresponding frequency distributions and characterizes the extent of self-similar spatial clustering. Formally, by rewriting Eq. (2.12) as:

$$N(\delta) = u\delta^{-D_c}, \tag{2.14}$$

where u is a constant, $N(\delta)$ gives the number of cells of size δ occupied by at least a single cluster, and it is obviously that $s = D_c$.

In estimating both s and D_c, a cell is considered as occupied regardless of the number of points n_i it contains. Thus, additional information can be obtained if one counts n_i as well and assign a weight to each box/cell such that those boxes with a larger number of points counts more as compared with those containing a smaller number of points. For this purpose, we define first the relative frequencies f_i such that $f_i = n_i/N$, where N is the total number of points in the structure, and $\sum_{i=1}^{N(\delta)} f_i = 1$, where $N(\delta)$ is the number of occupied boxes of size δ. This immediately leads to the definition of the Shannon entropy at scale δ:

$$\mathscr{S}(\delta) = -\sum_{i=1}^{N(\delta)} f_i \log f_i. \tag{2.15}$$

By writing again Eq. (2.13) as $\mathscr{S}(\delta) = u\delta^{-D_i}$, and using similar arguments as in deriving Eq. (2.13), one obtains:

$$D_i = \lim_{\delta \to 0} \frac{\mathscr{S}(\delta)}{\log(1/\delta)}. \tag{2.16}$$

In the case of a uniform distribution, all f_i are equal to $1/N(\delta)$, and thus Eq. (2.16) gives $\mathscr{S}(\delta) = \log N(\delta)$, and which corresponds to the maximum value of $\mathscr{S}(\delta)$. Therefore, in such cases, the box counting dimension is equivalent with the information dimension D_i, and thus $s = D_i$. However, in the case of non-uniform patterns, $S(\delta)$ reflects the non-uniformity of the distribution by giving less weight to the boxes with relatively low-number of points, and therefore $s > D_i$.

Spatial characterization of clustering can be effectively performed by considering the correlation function $C(\delta)$ of a point pattern [9], which is given by:

$$C(\delta) = \frac{1}{N} \sum_{i=1}^{N} C_i(\delta), \tag{2.17}$$

where N is the number of points in the pattern, and the quantity:

$$C_i \equiv \frac{1}{N} \sum_{i \neq j=1}^{N} \theta(\delta - r_{i,j}), \tag{2.18}$$

represents the number of distinct values of the distances $r_{i,j}$ between ith and jth point found inside a circle of radius δ centered on a given point in the pattern, and θ is the Heaviside function. For a non-uniform distribution it can be shown that the correlation function scales with δ according to $C(\delta) = u\delta^{D_{corr}}$, and which gives the probability that the distance between pairs of randomly selected points in the pattern is smaller than δ. Therefore, one can write:

$$D_{corr} \simeq \lim_{\delta \leftarrow 0} \frac{\log C(\delta)}{-\log(1/\delta)}, \tag{2.19}$$

which leads to $s > D_i > D_{corr}$, while in the case of uniform distributions $s = D_i = D_{corr}$.

In the next section we shall see that by generalizing the entropy given in Eq. (2.15), one can obtain a generalized spectrum of dimensions which relates the box-counting dimension s, information dimension D_i and correlation dimensions of various orders. Furthermore, in this framework, the shape of the dimension spectrum (and of other multifractal spectra) can be used to distinguish between non-uniform (i.e. fractal) and uniform (i.e. non-fractal) point distributions.

2.2.4 Mass and Surface Fractal Dimensions

A distinct set of fractal dimensions, such as perimeter, mass, or surface dimensions involve characterization of area-perimeter relationships as a function of their boundaries or of their plane/space-filling properties. In the following we shall deal with mass and surface fractal dimensions since they provide a deeper insight into the formation of fractal aggregates as well as on their surface characteristics [10, 11]. In addition, mass and surface fractal dimensions can be obtained experimentally at nano- and micro-scales using SAS methods [10–13], and they can be seen as complementary techniques to image analysis, but with the advantage that in SAS the number of investigated clusters is macroscopic, thus providing a much better statistics of the values of the fractal dimensions.

In SAS the yard-stick used in measuring the lengths in the box-counting method is now replaced by the x-ray or neutron wavelength. The working principles of this technique will be outlined in the next chapter, but for now it suffices to mention that in using SAS, we calculate in fact the fractal dimension of linear, planar or volumetric objects, randomly oriented and placed in the irradiated volume.

For this purpose, we consider a fractal of size l, consisting or regular structures, let's say balls of radius a whose sizes are such that $a \ll l$. Then the total number of balls N needed to cover the fractal can be written as [11, 14]:

$$N(r) \propto (l/a)^{D_m}, \tag{2.20}$$

where D_m is the mass fractal dimension. Thus, the number of balls enclosed by an imaginary sphere or radius r with a ball in the center is $N(r) \propto (r/a)^{D_m} \propto r^{D_m}$, which is valid within the fractal region $l_{min} \lesssim r \lesssim l$. Here, l_{min} is the minimal distance between the ball centers. By replacing the number of balls with their corresponding mass, Eq. (2.20) can be rewritten as:

$$M(r) \propto r^{D_m}, \tag{2.21}$$

and in this case, it gives the total "mass" of the fractal enclosed within a ball of radius r. Therefore, the quantity D_m is referred to as the mass fractal dimension. However, generally any other scalar quantity can be attached to the fractal support, such as volume, surface area etc. For structures embedded in a d-dimensional Euclidean space we have $0 < D_m < 3$. The higher the value of D_m the closer the structure.

By using Eq. (2.21) and considering that the volume of a ball inside the fractal is proportional to r^3, the density of the fractal becomes:

$$\rho(r) \propto r^{D_m - 3},\tag{2.22}$$

which shows that it is no longer constant but it decreases as the size of the volume increases. Comparing Eqs. (2.20) and (2.21) one can see that formally the mass fractal dimension coincides with the box-counting dimension.

For objects possessing a rough surface, it is useful to consider the surface area $S(r)$ found within the grid box of characteristic size r^2. Since the surface is rough, the number of boxes $N(r)$ that at least partially overlap the boundary, depends on r, and is proportional to r^{-D_s}, where D_s is called the surface fractal dimension. Thus, the surface are $S(r)$ can be regarded as equal to $r^2 N(r)$, and thus [11]:

$$S(r) \propto r^{2 - D_s}.\tag{2.23}$$

The value of D_s is between $d - 1$ and d, where d is the Euclidean dimension of the space in which the fractal is embedded. For $d = 3$, the surface fractal dimension $D_s \to 3$ when the surface is so folded that it almost fulfill all the available space, while for $D_s \to 2$ is very smooth and has an almost planar structure.

Recall that all natural systems contains a certain degree of polydispersity (either in shape or in size), and thus an important advantage of SAS over image analysis is that it provides values of the fractal dimensions averaged over a macroscopic volume. Since in a scattering experiment the phase information is lost and so we can not uniquely determine the structure, a combined use of both image analysis and SAS is a powerful approach in determining fractal dimensions and for elucidating fractal morphologies at nano- and micro-scales.

2.3 Multifractals

The physical meaning of the fractal dimension s introduced in the previous section is that it gives a description of how the density inside the fractal varies with respect to the length scale. Most of the existing theoretical fractal models are *homogeneous*, that is, they consist of a pattern which is repeated at every scale, and thus the fractal dimension is the same on all scales. However, for physical objects it is quite seldom that a single identical pattern repeats itself on all scales. Usually the self-similarity property can vary from point to point, i.e. these fractal objects are *heterogeneous*. This non-uniformity leads to the existence of objects having different fractal dimensions at different scales, and therefore a continuous spectra of dimensions is needed to describe them.

2.3.1 Formal Definition: Hausdorff Multifractal Spectra

A mass distribution μ may be spread over a region in such a way that the concentration of mass is very irregular. In particular, the set of points where the local mass concentration obeys a power-law, i.e. $\mu(B(x, r)) \simeq r^{\alpha}$, with $\alpha \geq 0$ is an index, and $r \to 0$, may give rise to different fractals, depending on α. Therefore a single measure may give rise to a large number of fractals. The measure μ together with such a rich structure is known as a *multifractal*, and the main aim of a multifractal analysis is to determine the structure of such fractals and their inter-relationship. According to Ref.[1], there are two basic approaches for multifractal analysis. In the first one—called *fine theory*, we determine the structure and the fractal dimensions defined by the local intensity of μ, while in the second one—called *coarse theory*, we consider the non-uniformity of distribution of μ over an r-mesh cubes, and then we take the limit $r \to 0$. In the first case the structure is described by a *singularity spectrum*, which is more convenient for theoretical developments, while in the second case, by a *coarse spectrum*, which is more convenient for practical implementations. Note that $r - mesh\ cubes$ on $\mathbb{R}^n$ are the cubes C of the form $[m_1 r, (m_1 + 1)r] \times \cdots \times [m_n r, (m_n + 1)r]$, with $m_1, \ldots, m_n \in \mathbb{Z}^*$.

Let's consider a measure μ on $\mathbb{R}^n$ such that $0 < \mu(\mathbb{R}^n) < \infty$. Then, the *Hölder exponent* (or *local dimension*) of the measure μ at a point $x \in \mathbb{R}^n$ is defined by [15]:

$$\dim_{\text{loc}}\mu(x) = \lim_{r \to 0} \frac{\log \mu(B(x, r))}{\log r}, \qquad (2.24)$$

where $B(x, r)$ is a ball of radius r centered on x. Then, the *fine (Hausdorff) multifractal spectrum* or the *singularity spectrum* of the measure μ is defined by [1]:

$$f_{\text{H}}(\alpha) = \dim_{\text{H}} \left\{ x \in \mathbb{R}^n : \dim_{\text{loc}}\mu(x) = \alpha \right\}. \qquad (2.25)$$

It is clear that $\forall \alpha \geq 0$ we have $0 \leq f_{\text{H}} \leq \dim_{\text{H}}(\text{spt}\mu)$.

For a coarse multifractal analysis, we consider again the measure μ in $\mathbb{R}^n$ with $0 < \mu(\mathbb{R}^n) < \infty$, and then we count the number of cubes C such tat $\mu(C) \propto r^{\alpha}$. For $\alpha \geq 0$ we write [1]:

$$N_r(\alpha) = \#\{r - \text{mesh cubes } C \text{ with } \mu(C) \geq r^{\alpha}\}, \qquad (2.26)$$

where the symbol # stands for "the number of". Then, the *coarse singularity spectrum* or the *coarse spectrum* of μ is given by [1]:

$$f_C(\alpha) = \lim_{\varepsilon \to 0} \lim_{r \to 0} \frac{\max\{0, \log(N_r(\alpha + \varepsilon)) - N_r(\alpha - \varepsilon)\}}{-\log r}, \qquad (2.27)$$

provided both limits exist. Thus, $f_C(\alpha)$ gives a global behavior of the fluctuations of the measure μ at scale r, but does not provide any information concerning the limiting behavior of μ in any point. For self-similar measures, and when $f_C(\alpha)$ exists, both approaches lead to the same multifractal spectra, and thus $\forall \alpha \geq 0$, we have $f_H(\alpha) = f_C(\alpha)$.

2.3.2 Practical Definition: Moment Method

We are concerned here with fractals on which a measure μ is distributed with constant density since for such objects multifractality manifests purely in the scaling properties of the geometry. Thus, the measure μ lying upon the fractal sets is simply the mass of the fractal and the multifractal spectra characterize the support itself.

As is the case of box-counting dimension, here let's consider first a uniform square grid of boxes with sizes l, and then define μ_i the proportion of the total mass of the fractal, inside the ith box. It is clear that for empty boxes, $\mu_i = 0$.

A central quantity of interest, which form the basis of any detailed investigations of multifractal properties is the *partition function* or the $q - th$ *moment* $Z_q(l)$ defined as [9]:

$$Z_q(l) = \sum_{i=1}^{N} \mu_i^q(l), \tag{2.28}$$

that is, Z_q is the sum over all boxes needed to cover the support. Here, $N \propto 1/l^2$ is the number of boxes of size $l \times l$ forming the grid. By assuming a power-law behavior of the partition function in the limit $l \to 0$ (i.e. $N \to \infty$), we can rewrite Eq. (2.28) as:

$$Z_q(l) \propto l^{(q-1)D_q}, \tag{2.29}$$

where D_q is called *order q generalized dimension*, and the factor $q - 1$ in the exponent ensures the normalization $Z_1(l) \equiv 1$ is satisfied.

Inside each box of size $l \times l$ the contribution of the measure μ increases with the size l is defined by the asymptotic relation according to:

$$\mu_i \propto l^{\alpha_i}, \tag{2.30}$$

where $\alpha_i = \alpha_i(l)$ is called the *crowding index* and is generally position dependent. At fixed l, the number of boxes with a given crowding index α can be written as:

$$N_\alpha(l) \propto l^{-f(\alpha)}, \tag{2.31}$$

that is, these boxes cover a subset having the fractal dimension $f(\alpha)$. It is clear from Eqs. (2.28) and (2.31) that $f(0)$ is the fractal dimension D_0 of the support, since $Z_0(l) = N_0(l)$. Also, since a multifractal is the union of all sub-fractals with dimensions $f(\alpha)$, it turns out that $f(0) > f(\alpha)$, $\forall \alpha$. This result is based on the property that in the case of composite fractals, that is fractals which are unions of fractal subsets, each with dimension $D_0^{(k)}$, the fractal dimension of the total set is given by the largest dimension of the subsets.

By using Eqs. (2.30) and (2.31) inside Eq. (2.28), one can rewrite the partition function such as:

$$Z_q(l) \propto \int l^{\alpha q - f(\alpha)} d\alpha. \tag{2.32}$$

In the limit $l \to 0$ the integral is dominated by the value of α which makes the exponent minimal. When $f(\alpha)$ and D_q are differentiable functions, this property leads to:

$$f(\alpha) = \alpha q - (q - 1)D_q, \tag{2.33}$$

where α is given by:

$$\alpha = \alpha(q) = \frac{d}{dq}(q - 1)D_q. \tag{2.34}$$

This shows that the spectrum of generalized dimension D_q and the spectrum of singularities $f(\alpha)$ are connected by a Legendre transform. The second term in Eq. (2.33) is usually denoted as:

$$\tau(q) \equiv (1 - q)D_q = \lim_{l \to 0} \frac{\log Z_q(l)}{-\log l}, \tag{2.35}$$

and is called the *scaling function*. Therefore, in terms of the partition function $Z_q(l)$ (see Eq. (2.28)) the generalized dimensions D_q can be written as:

$$D_q = \frac{1}{1 - q} \lim_{l \to 0} \frac{\log Z_q(l)}{-\log l} = \frac{1}{1 - q} \lim_{l \to 0} \frac{\log \sum_{i=1}^{N} \mu_i^q(l)}{-\log l}. \tag{2.36}$$

Since the measure μ is the relative mass of the ith box, we can write it explicitly as:

$$\mu_i \equiv \frac{M_i(l)}{M}, \tag{2.37}$$

where M_i is the mass of the ith box, and M is the total mass. When $q = 0$, one recover the box-counting dimension given by Eq. (2.13), since Eq. (2.36) gives:

$$D_0 = \lim_{l \to 0} \frac{\log N(l)}{-\log l}, \tag{2.38}$$

with $s \equiv D_0$ and $l = \delta$, and $N(l)$ being the number of boxes in the minimal cover. When $q = 1$, after applying L'Hopital's rule, one obtains:

$$D_1 = \lim_{l \to 0} \frac{\sum_{i=1}^{N} \mu_i \log \mu_i}{-\log l}, \tag{2.39}$$

which is called *information dimension*, it is related to Shannon's entropy and measures how the information scales with $1/l$ or how even is the data density. The higher the values of fractal dimension D_1, the more uniform density. When $q = 2$ one obtains D_2, which is called the *two-point correlation dimension* and is a measure of the correlation between pairs of points in each box and captures how the data is scattered, with higher compactness for increasing values of D_2. The other generalized dimensions indicate the correlations between triples (D_3), quadruples (D_4), etc. of points in each box.

There are however some practical limitations concerning the estimation of the generalized dimension spectrum D_q, which arise mainly from the non-uniformity and the porosity (lacunarity) of fractals, and which manifests themselves in poor sampling statistics for large and small values of α, and respectively in the presence of intrinsic oscillations on a double logarithmic scale of D_q. To avoid these inconveniences, several numerical procedures have been proposed, and which can be divided into two main classes: *fixed-size* methods, which are well suited for computing D_q for $q \geq 0$, and *fixed-mass* methods, which are best suited at $q < 0$, since they can overcome statistical fluctuations. However, both methods require a Legendre transformation to calculate the $f(\alpha)$ spectrum from D_q, as shown in Eq. (2.33). But, in turn, this requires first a smoothing of D_q and only then performing the actual transformation. This additional step eliminates the possibility to observe any singularities in the $f(\alpha)$ spectrum or in the scaling function $\tau(q)$, and thus it leads to the possibility of missing important physical phenomena in the scaling properties.

Several methods have been suggested to avoid, at least partially, these difficulties. The most common one consists in calculating first the derivative of the partition function, that is [16]:

$$\frac{d}{dq} Z_q(l) = \sum_{i=1}^{N} \mu_i^q(l) \log \mu_i(l), \qquad (2.40)$$

and then, by using Eqs. (2.29) and (2.33) the $\alpha(q)$ spectrum is rewritten as:

$$f(\alpha) = \lim_{l \to 0} \frac{\sum_{i=1}^{N} \tilde{\mu}_i(q, l) \log \tilde{\mu}_i(q, l)}{\log l}, \qquad (2.41)$$

where the normalized mass at each ith box is given by:

$$\tilde{\mu}_i(q, l) \equiv \mu_i^q(l) / \sum_{j=1}^{N} \mu_j^q(l). \qquad (2.42)$$

Furthermore, by using Eq. (2.34), one finally obtains:

$$\alpha(q) = \lim_{l \to 0} \frac{\sum_{i=1}^{N} \tilde{\mu}_i(q, l) \log \mu_i(l)}{\log l}. \qquad (2.43)$$

It was shown that $f(\alpha)$ spectrum given by Eq. (2.41) coincide with the Hausdorff dimension of measure theoretic support $\tilde{\mu}(q)$, while $\alpha(q)$ in Eq. (2.43) is an average singularity strength.

Although there are various methods such as histogram method, multifractal detrended fluctuation analysis or wavelet transform modulus maxima for studying the multifractality of particular types of objects, the moment method is widely used for the simplicity of its implementation and general applicability. It is particularly well suited for spatial data analysis or real space analysis of images acquired by various methods including atomic force microscopy, scanning electron microscopy, computed tomography etc. For practical purposes we list below the main steps in determining any of the multifractal spectra $(D_q, \tau(q), f(\alpha)$ or $\alpha(q))$ discussed above, with a focus on interpretation of the results:

1. The total number M of non-white pixels are counted.
2. The image is covered with a grid of squares of varying edge size l and the number of non-white pixels $M_i(l)$ in each box are counted.
3. The relative mass of the ith box is calculated from Eq. (2.37).
4. The generalized fractal dimensions spectrum D_q is calculated from Eq. (2.36). A sufficient range for q-values requires that both horizontal asymptotic values, $\alpha_{min} \equiv \lim_{q \to \infty} D_q$, that is the high-$q$ asymptotic value, and respectively $\alpha_{max} \equiv \lim_{q \to -\infty} D_q$, that is the low-$q$ asymptotic value, are reached by D_q. The interpretation of the D_q spectra is as follows: If the investigated object is a multifractal, then D_q *versus* q is monotonically a decreasing function with α_{min} and α_{max} characterizing the scaling properties of the most dense, and respectively of the most rarefied regions. If the object is not a multifractal, then D_q is a straight line with either integer or non-integer values. If D_0 is a non-integer then the object is a simple fractal, with fractal dimension D_0, while if D_0 is an integer the investigated object has a regular Euclidean shape, with $D_0 = 1$ for curves, and $D_0 = 2$ for surfaces.
5. If we are interested in determining also the scaling function $\tau(q)$, we additionally compute the partition function Z_q given by Eq. (2.28), for various values of l. For q-values we can select the same range as in computing D_q spectrum. The higher the values of q in Eq. (2.28), the more dense the selected regions.
6. If the plots of $Z_q(l)$ *versus* l, for selected values of q, are straight lines on a double logarithmic scale, then $\tau(q)$ is the slope corresponding to the exponent q. $\tau(q)$ is a strictly convex function and the line asymptotic to $\tau(q)$ at $q \to -\infty$ has the slope $-\alpha_{min}$, while the line asymptotic to $\tau(q)$ at $q \to \infty$ has the slope $-\alpha_{max}$. At $q = 1$, we shall have $\tau(1) = 0$, as indicated by Eq. (2.35).
7. For calculating the $f(\alpha)$ and $\alpha(q)$ spectra, we need to calculate the normalized mass according to Eq. (2.42).
8. $f(\alpha)$ spectrum is calculated according to Eq. (2.41). Since $f(\alpha)$ is the Legendre transform of $\tau(q)$, this implies that $f(\alpha)$ is continuous on $[\alpha_{min}, \alpha_{max}]$ and $f(\alpha_{min}) = f(\alpha_{max}) = 0$. Thus, $f(\alpha)$ is a single humped function with its maximum at D_0. For $q = 1$, and since $\tau(1) = 0$, it turns out from Eq. (2.33) that $f(\alpha) = \alpha$. Additionally, the derivative of $f(\alpha) - \alpha$ with respect to α is equal to

$q - 1$, which becomes 0 at $q = 1$. Thus, the $f(\alpha)$ spectrum shall be tangent to the line $f(\alpha) = \alpha$ at $q = 1$.

9. $\alpha(q)$ spectrum is determined from Eq. (2.43).

An important application is the possibility to develop a thermodynamic formalism for fractals and multifractals, where q and $\tau(q)$ play the role of inverse temperature and respectively the free energy. From the Legendre transform given by Eq. (2.33) the conjugate variables of q and $\tau(q)$ are α and $f(\alpha)$, the later ones thus playing the role of energy and entropy. For example, it is well known already that the Renyi dimensions proposed initially to describe the properties of strange attractors, are related to the $f(\alpha)$ spectrum of singularities of the corresponding measure. Thus the language of classical statistical mechanics provides a powerful theoretical framework that can be used for characterization of multifractals.

2.4 Methods for Constructing Fractal Structures

We describe several methods for constructing fractals and multifractals that have the potential of generating a large number of structures. The methods include affine transformations, iterative processes, and respectively a generalization of Brownian motion. We present their definition, main properties and few illustrative examples.

2.4.1 Iterated Function Systems

The method of iterated function systems (IFS) was introduced by Barnsley [4] and it provides a useful framework for generation of fractals, since the coordinates of the points forming the fractal, can be directly used to calculate the pair distance distribution function, and then the scattering intensity (see Chap. 3). By definition, an IFS consists of a complete metric space $(\mathbf{S}, d)$ together with a finite set of contraction mappings $w_n : \mathbf{S} \to \mathbf{S}$, with respective contractivity factors $s_n, n = 1, 2, \ldots, N$ [4]. Thus, an IFS is denoted as $\{\mathbf{S}; \ w_n, n = 1, 2, \ldots, N\}$ and $s = \max \{s_n, n = 1, 2, \ldots, N\}$. Generally, the transformation $f : \mathbf{X} \to \mathbf{X}$ on $(\mathbf{S}, d)$ is a contraction mapping if there exists a contractivity factor $0 \leq s < 1$, satisfying Eq. (2.10).

In the following we shall make use of the property that by considering an IFS with contractivity factor s, and denoting by $(\mathcal{H}(\mathbf{S}), h(d))$ the space of nonempty compact subsets with the Hausdorff metric $h(d)$, then the transformation $W : \mathcal{H}(\mathbf{S}) \to \mathcal{H}(\mathbf{S})$ defined according to [4]:

$$W(B) = \cup_{n=1}^{N} w_n(B), \ \forall B \in \mathcal{H}(\mathbf{S}), \tag{2.44}$$

is a contraction mapping on the complete metric space $(\mathcal{H}(\mathbf{S}), h(d))$ with:

$$h(W(B), W(C)) \leq s \cdot h(B, C) \,\forall B, \ C \in \mathcal{H}(\mathbf{X}). \tag{2.45}$$

The corresponding unique fixed point $A \in \mathcal{H}(\mathbf{S})$ is given by $A = \lim_{m \to \infty} W^{\circ m}(B)$ for any $B \in \mathcal{H}(\mathbf{S})$, and is called the *attractor* (or deterministic fractal) of the IFS [4]. It also satisfies the condition $A = \cup_{n=1}^{N} w_n(A)$.

For the purposes of our structural investigations, we shall use a deterministic algorithm, by starting with computation of a sequence of sets $\{A_m = W^{\circ m}(A)\}$, where the initial set A_0 is given. For simplicity, we restrict ourselves to IFS of the form $\{\mathbb{R}^2; \, w_n, n = 1, 2, \ldots, N\}$, where each mapping is an affine transformation. In the case of a random iteration algorithm, the calculations are based on ergodic theory.

Thus, in the deterministic case, we choose a compact set $A_0 \subset \mathbb{R}^2$, and then the quantities A_m are determined, such as [4]:

$$A_m = \cup_{n=1}^{N} w_n(A_{m-1}), \ for \ m = 1, 2, \ldots. \tag{2.46}$$

This corresponds to building the sequence $\{A_m : m = 0, 1, \ldots\} \subset \mathcal{H}(\mathbf{S})$, which converges to the attractor of the IFS.

For the random iteration algorithm, we assign first the probability $p_n > 0$ to w_n for $n = 1, 2, \ldots, N$, where $\sum_{n=1}^{N} p_n = 1$. Then a point $x_0 \in \mathbf{X}$ is chosen, which is followed by calculation of the other points, according to:

$$x_k \in \{w_1(x_{k-1}), w_2(x_{k-1}) \ldots, w_N(x_{k-1})\}. \tag{2.47}$$

Here p_n is the probability of the event $x_k = w_n(x_{k-1})$, and $k = 1, 2, \ldots$. This procedure generates the sequence $\{x_k : k = 0, 1, \ldots\} \subset \mathbf{S}$ which also converges to the attractor of IFS.

As an application of the previous theory, we consider here a Sierpinski triangle of edge length $a = 1$. We choose also a system of Cartesian coordinated centered in the origin, and with ox-axis parallel to one of the edges. Thus we have $N = 3$, and the matrix representation of the IFS is given by

$$w_i \begin{bmatrix} x \\ y \end{bmatrix} = \begin{bmatrix} a_i & b_i \\ c_i & d_i \end{bmatrix} \begin{bmatrix} x \\ y \end{bmatrix} + \begin{bmatrix} e_i \\ f_i \end{bmatrix} \tag{2.48}$$

where the coefficients a_i, b_i, c_i, d_i, e_i, f_i with $i = 1, 2, 3$ are given by: $w_i = \{1, 2, 3\}$, $a_i = \{1/2, 1/2, 1/2\}$, $b_i = \{0, 0, 0\}$, $d_i = \{1/2, 1/2, 1/2\}$, $e_i = \{0, -1/4, 1/4\}$, $f_i = \{1/(4\sqrt{3}), -1/(4\sqrt{3}), -1/(4\sqrt{3})\}$. In Fig. 2.1 are shown the points (black) given by deterministic algorithm defined by Eq. (2.48) for $m = 1, 2, 3$ and 4.

For the random iteration algorithm, there exist several ways of introducing randomness. Here, we generate the SG by "playing" the chaos game [4] on three vertices which do not all lie on a line. In this approach, the points of the SG are obtained starting with an initial point chosen at random, and calculating each subsequent point as a fraction (here $\beta_s = 1/2$) of the distance between the previous point and one of

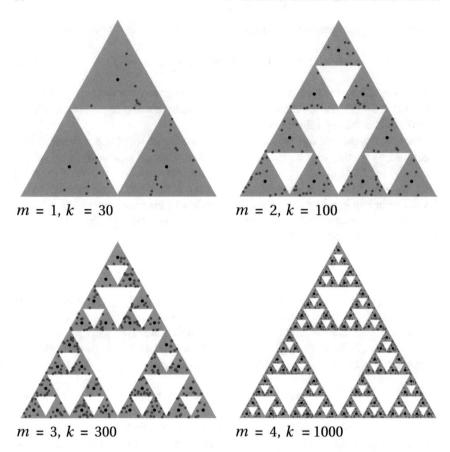

$m = 1, k = 30$ $m = 2, k = 100$

$m = 3, k = 300$ $m = 4, k = 1000$

Fig. 2.1 Superposition of SG obtained from the deterministic (black points) and random (blue points) algorithms (see Ref. [17]). Here, m is th fractal iteration number and k is the number of points (blue) used to generate the SG

the vertices (selected randomly at each iteration) of the triangle. By repeating this procedure for a large number of points, and selecting the vertex at random on each iteration, the SG is obtained. Thus, points over the attractor in random order are generated with equal probabilities $p = 1/3$. This is opposed to other methods which test each pixel to see whether it belongs to the fractal [4]. Blue points in Fig. 2.1, shows the result of chaos game for various values of the number of points k. The main feature shown is that by increasing k, leads to a better agreement with SG. One can notice that a value of $k = 1000$ is quite sufficient to generate a SG at iteration $m = 4$. The comparison performed in Fig. 2.1 clearly indicates that, excepting few points which were omitted, all the others belong to the SG. In this way, a direct comparison between their respective structure factors can be performed (see next section).

2.4.2 Cellular Automata

Cellular automata (CA) are built on a regular lattice of cells where the state of each cell is updated at discrete steps [18, 19]. This is performed according to a rule which depends on the state of cells in its neighborhood. Formally, CA is a tuple

$$\mathbb{A} = (d, S, N, \Phi) \tag{2.49}$$

where $d \in \mathbb{Z}^*$ indicates the dimension of CA, $S = \{0, 1, 2, \ldots M - 1\}$ represent a finite set of states, $N = \{r_1, r_2, \ldots, r_l\} \in (\mathbb{Z}^d)^l$ is a vector containing the n cells from the neighborhood, and $\Phi : S^n \to S$ is a local rule of evolution. There exists 4 main classes of behaviour for each CA pattern: (a) disappear with time; (b) evolve at a fixed finite size; (c) grow indefinitely at a fixed speed; (d) grow and contract irregularly. We focus here on type (c) since their limiting form yields a chaotic aperiodic behavior, and often contain fractal structures.

It is useful that in SAS analysis of CA [20, 21] to consider first an $M = p^t$-state CA, where p is a *prime* and $t \in \mathbb{N}$. Then, the value a_i^k of a cell at position i and step k is determined from a transition rule and depends on the states of its neighborhood $i + r_1, i + r_2, \ldots i + r_l$ at step $k - 1$. Therefore:

$$a_i^k = \phi\left(a_{i+r_1}^{k-1}, a_{i+r_2}^{k-1}, \ldots, a_{i+r_l}^{k-1}\right), \tag{2.50}$$

where $(r_1, r_2, \ldots, r_l) \in (\mathbb{Z}^d)^l$ is called a neighborhood index, and $d \in \mathbb{N}$ is the dimensionality of the lattice. In particular, a one-dimensional $(d = 1)$ additive CA (ACA) is described by a linear rule of the type

$$a_i^k = c_1 a_{i+r_1}^{k-1} + c_2 a_{i+r_2}^{k-1} + \cdots + c_l a_{i+r_l}^{k-1} \quad \text{mod } M, \tag{2.51}$$

where the coefficients $c_i \in \mathbb{N}$ and $i = 1, \ldots, l$. Here we consider that at $k = 0$ the initial row contains only one occupied site with value 1. At $k = 1$, Eq. (2.51) generates another row based on the configuration at $k = 0$. By repeating the same procedure for each subsequent step k, a new row is added resulting in a pattern embedded in a two-dimensional Euclidean space (Fig. 2.2). Thus, the evolution of an ACA is done only with a single line of cells at every step.

Two well-known recurrence relations for the M-state ACA can be obtained by using Eq. (2.51). In the first case we have $r_1 = -1, r_2 = 1$ and all other indices equal to zero, while $c_1 = c_2 = 1$ and all other coefficients are equal to zero. Therefore:

$$a_i^k = a_{i-1}^{k-1} + a_{i+1}^{k-1} \quad \text{mod } M. \tag{2.52}$$

When the values $r_1 = -1, r_3 = 1$ are chosen, while all other indices equal to zero, with $c_1 = c_2 = c_3 = 1$ and all other coefficients equal to zero, the recurrence relation can be written as:

$$a_i^k = a_{i-1}^{k-1} + a_i^{k-1} + a_{i+1}^{k-1} \quad \text{mod } M. \tag{2.53}$$

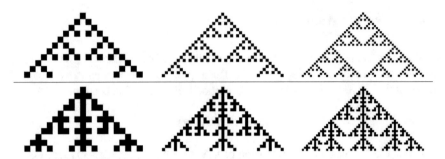

Fig. 2.2 Structural evolution of 2-state ACA at $k = 10$, 20 and 30 steps. Upper row: Rule 90. Lower row: Rule 150 (see Ref. [21])

As an example, we illustrate here two representative cases of ACA, when $M = 2$ and 3. In the first case, Eqs. (2.52) and (2.53) represent, what is known in the literature as Rule 90 and respectively Rule 150.

Figure 2.2 shows the evolution of Rule 90 (left side) and Rule 150 (right side) at various discrete steps k. Rule 90 shows that for $k = 8$ and $k = 16$ steps, the generated patterns are exactly self-similar and they correspond to a Sierpinski triangle at 1-th and respectively 2-nd iteration. For other values of steps, such as for $k = 28$ one obtains a non-complete fractal structure [20, 21]. Rule 150 also generates self-similar fractal structure in the sense that it can be decomposed into self-similar subparts.

A more general case of fractal structure can be obtained by keeping the same value for the indices and for the coefficients used to generate Rule 90 and Rule 150, but considering that the states of each site can take arbitrarily values. We name these patterns "M-state Rule 90" and respectively "M-state Rule 150" in order to underline that the recurrence relations are the same for each number of possible states (see next Chapter).

Since the M-state ACA can generate self-similar structures, one can characterize them using the Hausdorff (fractal) dimension [2] as

$$D = \lim_{k \to \infty} \frac{\log N_k}{\log k}, \tag{2.54}$$

where N_k is the total number of non-zero state sites.

As an alternative method to calculate fractal dimensions for ACA, Wilson has suggested the transition matrix (TM) method [22]. In this approach one considers a set of blocks of length m with all possible configurations of M states. The difference in positions of the first and last neighbors in Eq. (2.51), i.e. $r_l - r_1 \leq m$, shall be smaller than the length of the blocks. Thus, the number of non-trivial blocks is $h = 2^m - 1$. The TM method shows how many blocks of a certain type are generated by the transition rule from hth block. The fractal dimension D of the ACA is then given by [22]:

$$D = \log_M \lambda, \tag{2.55}$$

where λ is the largest eigenvalue.

However, a detailed description of most physical phenomena occurring on fractals needs a characterization of the measures such as the distribution of concentration or pressure associated to each point of the support. As we saw in the previous section, an efficient method which allows us to characterize the heterogeneities of the measures involves multifractal measures and will be used in the following chapter.

References

1. Falconer, K.J.: Fractal Geometry: Mathematical Foundations and Applications. Wiley, New York (2003)
2. Hausdorff, F.: Mathematische Annalen **79**(1–2), 157 (1918)
3. Mandelbrot, B.B.: The Fractal Geometry of Nature. Freeman, W.H (1982)
4. Barnsley, M.F.: Fractals Everywhere. Academic, New York (1988)
5. Baumann, G.: In Mathematica® for Theoretical Physics, pp. 773–897. Springer, New York (2005)
6. Lynch, S.: In Dynamical Systems with Applications using Mathematica®, pp. 331–361. Birkhäuser, Boston
7. Newkome, G.R., Wang, P., Moorefield, C.N., Cho, T.J., Mohapatra, P.P., Li, S., Hwang, S.H., Lukoyanova, O., Echegoyen, L., Palagallo, J.A., Iancu, V., Hla, S.W.: Science (New York, N.Y.) **312**(5781), (1782) (2006)
8. Rathgeber, S., Pakula, T., Urban, V.: J. Chem. Phys. **121**(8), 3840 (2004). http://aip.scitation.org/doi/10.1063/1.1768516
9. Hentschel, H., Procaccia, I.: Phys. D: Nonlinear Phenom. **8**(3), 435 (1983)
10. Martin, J.E., Hurd, A.J.: J. Appl. Crystallogr. **20**(2), 61 (1987)
11. Schmidt, P.W.: J. Appl. Crystallogr. **24**(5), 414 (1991)
12. Bale, H.D., Schmidt, P.W.: Phys. Rev. Lett. **53**(6), 596 (1984)
13. Teixeira, J.: J. Appl. Crystallogr. **21**(6), 781 (1988)
14. Cherny, A.Y., Anitas, E.M., Osipov, V.A., Kuklin, A.I.: Phys. Rev. E **84**(3), 036203 (2011)
15. Evans, L.C.: Partial Differential Equations, 2nd edn. American Mathematical Society, Providence (2010)
16. Chhabra, A., Jensen, R.V.: Direct Determination of the f (c) Singularity Spectrum. Technical Report (1989)
17. Anitas, E.M., Slyamov, A.: PLOS ONE **12**, 1 (2017)
18. Wolfram, S.: Rev. Mod. Phys. **55**(3), 601 (1983)
19. Wolfram, S.: A New Kind of Science. Wolfram Media, Champaign (2002)
20. Anitas, E.M., Slyamov, A.: Ann. Phys. **530**(6), 1800004 (2018)
21. Anitas, E.M., Slyamov, A.M.: Ann. Phys. **530**(11), 1800187 (2018)
22. Willson, S.J.: Phys. D: Nonlinear Phenom. **24**(1–3), 179 (1987)

Chapter 3
Small-Angle Scattering Technique

Abstract This chapter provides a basic overview of SAXS and SANS techniques, which includes definitions of the basic notions, description of main properties of x-rays and neutrons, a general mathematical background which describes the physical processes in a scattering experiment, and main procedures in experimental data analysis and extraction of structural parameters. We also present derivation of form factors of basic Euclidean shapes which will form the basic units of the fractals introduced thereafter, and describe the main numerical methods for calculating the scattering intensity when analytic form factors are not available. Although, in principle SAS with light (SALS) and with fast electrons (SAES) is also possible, we intentionally omit their discussion here since SALS can not be used for non-transparent samples while SAES requires large wavelengths.

3.1 Introduction

A basic scattering event involves a beam of particles (x-ray/neutrons) hitting a target (sample) whose structure is to be determined from the distribution of scattered particles at different scattering angles θ (Fig. 3.1). While in a SAS experiment, x-rays are scattered elastically by the electron cloud, neutrons are scattered elastically, coherently or incoherently by nuclei or by the magnetic moments associated with unpaired electron spins in magnetic materials. X-rays are also scattered by nuclei but their mass is at least one thousand times greater than the mass of the electrons, thus the nuclear scattering energy is 10^6 smaller, and therefore nuclear scattering is neglected in x-ray scattering.

To describe the scattering process, let's consider a beam of particles with wavelength λ, propagating with the wave vector $\mathbf{k} = (2\pi/\lambda)\,\hat{\mathbf{k}}$ along the z-axis which coincides with the axis of the collimator with beam opening Δr. The role of the collimator is to provide a monochromatic and very well collimated beam, so that the energy spectrum to be enough sharp and concentrated at a unique eigenvalue of energy E. The number of particles of a particular type is recorded by the detector with a small opening $d\Omega$, situated at a large distance from the sample, and whose position $\mathbf{r} = (x, y, z)$ is along the direction $\hat{\mathbf{r}}$ at an angle (θ, ϕ) with respect to the

© The Author(s), under exclusive license to Springer Nature Switzerland AG 2019   33
E. M. Anitas, *Small-Angle Scattering (Neutrons, X-Rays, Light)*
from Complex Systems, SpringerBriefs in Physics,
https://doi.org/10.1007/978-3-030-26612-7_3

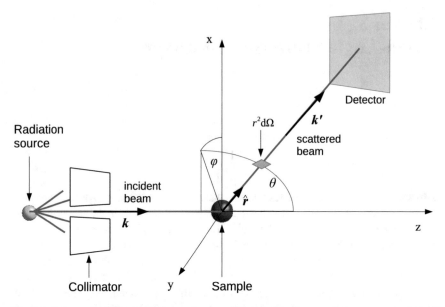

Fig. 3.1 Schematic setup of a SAS experiment, where a beam of x-rays or neutrons are released from a source. A small fraction of particles pass through the collimator, then they hit the sample, and the scattered ones are recorded by the detector. Here, **k** and **k**$'$ are the incident and, respectively the scattered wave vectors, $\hat{r}$ is the unit vector along scattered beam defined by angles ϕ and θ in the xyz Cartesian coordinate system, and Ω is the solid angle

direction of propagation of the incident beam. Therefore, at position detector, the scattered beam from the sample can be fully characterized by the wave vector **k**$'$ and energy E'.

The main characteristics which describe the scattering process are the scattering vector **q** defined by [1]:

$$\mathbf{q} = \mathbf{k}' - \mathbf{k},\tag{3.1}$$

and the energy transition $\hbar\omega = E' - E$. Since SAS structural investigations are performed in elastic scattering mode we have $E' = E$ and $|\mathbf{k}'| = |\mathbf{k}| = k$, and therefore the magnitude of the scattering vector is obtained from:

$$q = \sqrt{k^2 + k'^2 - 2kk'\cos\theta} = \sqrt{2k^2\left(1 - \cos\theta\right)}.\tag{3.2}$$

By using the property that $\cos\theta = 1 - 2\sin^2\theta/2$, and $k = 2\pi/\lambda$, one obtains:

$$q = \frac{4\pi}{\lambda}\sin\frac{\theta}{2}.\tag{3.3}$$

3.2 Basics of Scattering Theory

During the flight time of the particles from the collimator to detector, the occurring physical phenomena can be very complex. These includes absorption together with excitation of electron/nuclear system, followed by de-excitation and re-emission of a particle, charge and spin exchange, or particles may interact with each other. In addition, multiple scattering can occur, that is after the incident particles interact with all the nuclei and electrons and thus they become a source of secondary waves (first order / Born approximation), their superposition can be again scattered by all nuclei and electrons (second order approximation) and so on. However, calculation of these successive approximations in the framework of perturbation theory is a very difficult task, and usually one restricts only to the first approximation, which is valid only for weakly scattering centers or fields. In SAS, generally we neglect all these physical effects, and discuss the scattering problem by using the Schrödinger equation with an appropriate Hamiltonian of the particle-sample system and proper boundary conditions. This approach is very useful to get a deeper insight into the underlying physical phenomena.

Under these assumptions, the state of the particle-sample system can be described by a two-particle wave function $\Psi(\mathbf{r}_p, \mathbf{r}_s)$ and by a potential $V(\mathbf{r})$, where $\mathbf{r}_p$ and $\mathbf{r}_s$ are the position vectors of the particle and respectively of the sample, and $\mathbf{r} = \mathbf{r}_p - \mathbf{r}_s$ [2]. The potential field $V(\mathbf{r})$ acts in a finite range of space and represents the distribution of the electric charge density in the case of x-ray scattering, and the nuclear and spin density distribution in the case of neutron scattering. Since the nuclear scattering potential has a short range, and thus the particle is not under the influence of the sample neither when they are emitted from the collimator nor when they arrive at detector, the scattering process can be described in the framework of "s-wave scattering", and thus the restriction to the first order approximation in SAS as mentioned above, is a safely assumption. The same holds true also for x-ray scattering, which are described also by short order potentials.

In addition, since the considered system consists from the particle and the sample, then the motion of their center-of-mass $\mathbf{R}$ can be factored out, and since the potential field does not depends explicitly on $\mathbf{R}$, we can write the two-particle wave function as a product of two single-particle wave functions, such as: $\Psi(\mathbf{r}_p, \mathbf{r}_s) = \Phi(\mathbf{R})\psi(\mathbf{r})$. Furthermore, in a physical experiment the mass of the sample (m_s) is much heavier than the mass of the incident particle m_p and thus one can consider that the center-of-mass of the sample remains unchanged during the scattering process. Thus, the relative distance $\mathbf{r}$ between the sample and the incident particle is the laboratory frame of reference of the incident particle, while the inverse of the reduced mass $(m_p + m_s)/(m_p m_s)$ becomes the mass m of the incident particle. Under these additional assumptions, the most general description is provided by the time-dependent Schrödinger equation, given by:

$$i\hbar \frac{\partial \psi(\mathbf{r}, t)}{\partial t} = \left[-\frac{\hbar^2}{2m}\nabla^2 + V(\mathbf{r}) \right] \psi(\mathbf{r}, t), \tag{3.4}$$

since it involves discrete particles. The solution to this equation, in the region where
$V(\mathbf{r}) = 0$, is given by [2]:

$$\psi(\mathbf{r}, t) = \frac{1}{(2\pi)^3} \int d^3 k A(\mathbf{k}) \psi_k(\mathbf{r}, t), \tag{3.5}$$

where $\psi_k(\mathbf{r}, t) = \exp(i\mathbf{k} \cdot \mathbf{r}) \exp\left(-i\hbar k^2 t/(2m)\right)$ and $A(\mathbf{k})$ represents the probability
amplitude for finding the wave number $\mathbf{k}$ in the initial state. However, this solution
does not take into account the conditions that reflect the investigated physical phenomenon. In order to overcome this issue one may use the Huygens principle and
in the limit $r \to \infty$, for an arbitrarily wave vector $\mathbf{k}$ and time t, a single scattering
center give rise to the following wave function:

$$\psi_k(\mathbf{r}) = e^{i\mathbf{k}\cdot\mathbf{r}} + \frac{1}{r}e^{ikr}f_k(\theta, \phi), \tag{3.6}$$

where $f(\theta, \phi) \equiv f(\hat{\mathbf{r}})$ represents the scattering amplitude as a function of azimuthal
and planar scattering angles (see Fig. 3.1). Then, the scattering problem can be simplified by solving the stationary Schrödinger equation [2]:

$$\left[-\frac{\hbar^2}{2m}\nabla^2 + V(\mathbf{r})\right]\psi_k(\mathbf{r}) = E\psi_k(\mathbf{r}), \tag{3.7}$$

that is consistent with condition (3.6) of an incoming plane wave $\psi_k(\mathbf{r}) = e^{i\mathbf{k}\cdot\mathbf{r}}$ and an
outgoing spherical wave. For this purpose we introduce the Green function, satisfying
the following differential equation:

$$\left[-\frac{\hbar^2}{2m}\nabla^2 + E\right]G\left(\mathbf{r}, \mathbf{r}'|E\right) = \delta(\mathbf{r} - \mathbf{r}'), \tag{3.8}$$

where E is given by the energy of the incident wave, and $\mathbf{r}'$ is the position of the
source (Fig. 3.2). Thus, Eq. (3.7) becomes [2]:

$$\left[\frac{\hbar^2}{2m}\nabla^2 + E\right]\psi_k(\mathbf{r}) = V(\mathbf{r})\psi_k(\mathbf{r}). \tag{3.9}$$

The sought solution of Eq. (3.9) is a power series in powers of the interaction potential
V. For this purpose we transform Eq. (3.9) into an integral equation of the form [2]:

$$\psi'_k(\mathbf{r}) = \psi_k(\mathbf{r}) + \int_{\mathscr{V}} d^3 r' G\left(\mathbf{r}, \mathbf{r}'|E\right) V(\mathbf{r}')\psi'_k(\mathbf{r}'), \tag{3.10}$$

where $\mathscr{V}$ denotes the volume of the sample, and the product $V(\mathbf{r}')\psi'_k(\mathbf{r})$ is the
inhomogeneity of Eq. (3.8). This form is known in the literature as the Lippmann–
Schwinger equation. The index $\mathbf{k}$ in the function ψ' indicates that this state has

evolved from a previous state which was a plane wave of wave vector $\mathbf{k}$. Note that in the limit $V(\mathbf{r}) \to 0$ one recovers $\psi'_{\mathbf{k}}(\mathbf{r}) = \psi_{\mathbf{k}}(\mathbf{r})$, that is the scattered and incident waves coincide. The requirement of spherical outgoing waves $\psi'_{\mathbf{k}}(\mathbf{r})$ emerging from the sample, imposed by Eq. (3.6), allows us to write the Green function appearing in Eq. (3.8) as:

$$G\left(\mathbf{r}, \mathbf{r}'|E\right) = -\frac{2m}{\hbar^2} \frac{1}{4\pi} \frac{e^{ik|\mathbf{r}-\mathbf{r}'|}}{|\mathbf{r}-\mathbf{r}'|}, \tag{3.11}$$

which describes the stationary radiation generated at $\mathbf{r}'$ of a particle with energy E. Since in an experimental situation, the distance between the sample and the detector is always much bigger than the size of the scattering volume, that is we work in the Fraunhofer approximation with $r' \ll r$, one can safely assume that $|\mathbf{r} - \mathbf{r}'| \simeq r - \hat{\mathbf{r}} \cdot \mathbf{r}'$, where $\hat{\mathbf{r}} = \mathbf{r}/|\mathbf{r}|$. Therefore $1/|\mathbf{r} - \mathbf{r}'| = 1/r + \mathcal{O}\left(1/r^2\right)$, and the asymptotic form of Eq. (3.11) reads as:

$$G(\mathbf{r}, \mathbf{r}'|E) = -\frac{2m}{\hbar^2} \frac{1}{4\pi} \frac{e^{ikr}}{r} e^{-k\hat{\mathbf{r}} \cdot \mathbf{r}'} + \mathcal{O}(1/r^2). \tag{3.12}$$

By inserting Eq. (3.12) into Eq. (3.10) one recovers the wave function given by Eq. (3.6), and thus the solution of the stationary Schrödinger equation is an incident plane wave together with a spherical one multiplied by scattering amplitudes [2]:

$$f_{\mathbf{k}}(\hat{\mathbf{r}}) = -\frac{2m}{\hbar^2} \frac{1}{4\pi} \int d^3r' e^{-\mathbf{k}' \cdot \mathbf{r}'} V(\mathbf{r}') \psi'_{\mathbf{k}}(\mathbf{r}'), \tag{3.13}$$

which is independent of the distance between scattering volume and detector.

Equation (3.10) is generally solved by using an iterative procedure. Therefore, at zero-th order in powers of the interaction potential V, the scattering wave is given by $\psi'^{(0)}_{\mathbf{k}}(\mathbf{r}) = e^{i\mathbf{k}\cdot\mathbf{r}}$ which represents an incident plane wave. Higher iterations of Lippmann–Schwinger equation are obtained according to [2]:

$$\psi'^{(n+1)}_{\mathbf{k}}(\mathbf{r}) = \psi'^{(0)}_{\mathbf{k}}(\mathbf{r}) + \int d^3r' G(\mathbf{r}, \mathbf{r}'|E) V(\mathbf{r}') \psi'^{(n)}_{\mathbf{k}}(\mathbf{r}'), \tag{3.14}$$

and generates the wave functions in powers of the potential V, such as:

$$\psi'_{\mathbf{k}}(\mathbf{r}) = \psi'^{(0)}_{\mathbf{k}}(\mathbf{r}) + \psi'^{(1)}_{\mathbf{k}}(\mathbf{r}) + \psi'^{(2)}_{\mathbf{k}}(\mathbf{r}) + \cdots, \tag{3.15}$$

and which can be rewritten as:

$$\psi'_{\mathbf{k}}(\mathbf{r}) = \psi'^{(0)}_{\mathbf{k}}(\mathbf{r}) + G(\mathbf{r}, \mathbf{r}'|E) V(\mathbf{r}) \psi'^{(0)}_{\mathbf{k}}(\mathbf{r}) + \cdots. \tag{3.16}$$

The first term expresses single scattering events of the incident plane wave, the second one expresses scattering event of the first order, and so on. However, even for the first-order term the calculations become quite complicated, and therefore a common approach in SAS is to consider only the zero-th order. Thus, in this framework, the general solution of Eq. (3.13) is given by [2]:

$$f_{\mathbf{k}}^{(1)}(\hat{\mathbf{r}}) = -\frac{2m}{\hbar^2}\frac{1}{4\pi}\int d^3r' e^{-i\mathbf{k}'\cdot\mathbf{r}'}V(\mathbf{r}')e^{i\mathbf{k}\cdot\mathbf{r}'} = -\frac{2m}{\hbar^2}\frac{1}{4\pi}\mathscr{F}\{V(\mathbf{r})\} \qquad (3.17)$$

where the symbol $\mathscr{F}\{\cdots\}$ represents the Fourier transform.

3.3 Scattering Cross Sections and SAS Intensity

In a SAS experiment one measures the flux (current density) of a scattered beam normalized to the current density of incident particles as a function of the scattering angle. This variation is used thereafter for determination of the sample structure. Physically, the detector situated at distance r along the direction of $\hat{\mathbf{r}}$ at an angle $\Omega = (\theta, \phi)$ (see Fig. 3.1) counts the number of particles dN scattered in the solid angle $d\Omega = \sin\theta d\theta d\phi$ per unit time dt. Thus, the most important quantity is the differential scattering cross section $d\sigma/d\Omega$, which is defined by:

$$\frac{d\sigma}{d\Omega} = \frac{J_s}{J_i}r^2, \qquad (3.18)$$

where J_s and J_i are the current densities of incoming and respectively of the scattered beams. By using Eq. (3.6) one can see that the scattered wave has the expression $\psi_s = (1/r)exp(ikr)f(\Omega)$, which can be used to calculate the current density of the scattered beam according to [3]:

$$\mathbf{J}_s \equiv \frac{i\hbar}{2m}\left(\psi_s(\mathbf{r})\nabla\psi_s^*(\mathbf{r}) - \psi_s^*(\mathbf{r})\nabla\psi_s(\mathbf{r})\right). \qquad (3.19)$$

Therefore, one obtains the approximation:

$$\mathbf{J}_s(\mathbf{r}) \simeq J_i\frac{1}{r^2}|f(\Omega)|^2\hat{\mathbf{r}} + \mathcal{O}(\frac{1}{r^3}) + \cdots, \qquad (3.20)$$

which inserted back into Eq. (3.18) gives:

$$\frac{d\sigma}{d\Omega} = |f(\Omega)|^2. \qquad (3.21)$$

This relation connects the scattering theory (through the scattering amplitude) to the differential scattering cross section, which is a measurable experimental quantity. By performing integration over the solid angle Ω, one obtains the total scattering cross section [1]:

$$\sigma \equiv \int_0^{4\pi} \left(\frac{d\sigma}{d\Omega} \right) d\Omega = \int_0^{2\pi} \int_0^{\pi} \frac{d\sigma}{d\Omega} \sin \theta d\theta d\phi, \qquad (3.22)$$

which represents the total number of particles scattered in all directions per second, divided to the flux of the incident beam. However, physical samples may have different volumes and thus they show different scattering cross sections. In order to avoid this issue, the total differential scattering cross section is normalized to the volume V of the sample, thus giving the scattering intensity I as a function of the scattering wave vector q [1]:

$$I(q) \equiv \frac{1}{V} \frac{d\sigma}{d\Omega}. \qquad (3.23)$$

The scattering intensity has the dimension cm^{-1} $ster^{-1}$, and is of central importance in SAS, since virtually all the quantities describing the properties of the sample from SAS data, are expressed in terms of $I(q)$. Note that in the SAS literature, the dimensionless unit of the solid angle is omitted, and intensities are shown only in cm^{-1}.

3.4 Diffraction and the Array Theorem

A fundamental property of the scattering intensity given in Eq. (3.23) is that generally, it can be decomposed into a product of two functions which reflect the shape of the scattering objects (form factor), and respectively their distribution inside the scattering volume (structure factor). We can show this by making a close analogy with diffraction from a set of similarly oriented, identical diffraction apertures denoted here by Σ, containing N transparent regions, labeled by j, and where the summation over the amplitudes obtained from each single aperture has to be taken into account.

Let's consider first a two-dimensional diffracting aperture, laid in the (x, y) plane, illuminated in the positive z direction. In the observation plane (u, v), parallel to Σ, the complex-valued amplitude of the obtained diffraction image, computed using the framework of scalar theory of diffraction, according to the Huygens-Fresnel principle, reads as [4]:

$$A(u, v) = \frac{z}{i\lambda} \iint_{\Sigma} A(x, y) \frac{e^{ikr}}{r^2} dx \, dy. \qquad (3.24)$$

Here, $r = \sqrt{z^2 + (u - x)^2 + (v - y)^2}$ is the distance between two arbitrarily points taken, respectively, from the plane containing Σ, and from the observation plane. In the Fraunhofer diffraction this distance must satisfy the condition of being much bigger than the radiation wavelength λ. Performing a binomial expansion of the square root in Eq. (3.24), and keeping only the first two terms, gives [4]:

$$r \approx z \left(1 + \frac{(u-x)^2}{2z^2} + \frac{(v-y)^2}{2z^2} \right),$$ (3.25)

which can be used to write the Fresnel diffraction integral:

$$\frac{A(u,v)}{P(u,v)} = \int\int_{-\infty}^{+\infty} \left\{ A(x,y)e^{i\frac{k}{2z}(x^2+y^2)} \right\} e^{-i\frac{2\pi}{\lambda z}(ux+vy)} dx \, dy,$$ (3.26)

where and $k = 2\pi/\lambda$, and the quantity $P(u,v)$ is written as:

$$P(u,v) = \frac{e^{ikz}e^{i\frac{k}{2z}(u^2+v^2)}}{i\lambda z}.$$ (3.27)

By taking into account the additional condition: $z \gg k \, \mathrm{Max}(x^2 + y^2)/2$, we have $\exp(\frac{k}{2z}(x^2 + y^2)) \simeq 1$. Thus, Eq. (3.26) can be rewritten as:

$$A(u,v) = P(u,v) \int\int_{-\infty}^{+\infty} A(x,y)e^{-i\frac{2\pi}{\lambda z}(ux+vy)} dx \, dy.$$ (3.28)

Writing the spatial frequencies as $p = u/(\lambda z)$ and respectively $s = v/(\lambda z)$, and ignoring the multiplicative phase factor $P(u,v)$ preceding the integral in Eq. (3.28), the amplitude becomes the Fourier transform of the distribution of the Σ aperture. Since the illumination is considered to consist from a monochromatic, unit-amplitude plane-wave, at normal incidence, and that the field distribution across the aperture is equal to its transmission function $T(x,y)$, the frequency distribution of the diffraction amplitude in the phase space becomes:

$$A(p,s) = \int\int_{-\infty}^{+\infty} T(x,y)e^{-2i\pi(px+sy)} dx \, dy.$$ (3.29)

For a set of N identical apertures the well-known frequency distribution of the diffraction amplitude of a single aperture (Eq. (3.29)) can be rewritten as [5]:

$$A(p,s) = \sum_{j=1}^{N} \int\int_{-\infty}^{+\infty} T(x,y)e^{-2i\pi\left(p(x+x_j)+s(y+y_j)\right)} dx \, dy.$$ (3.30)

The position of a point in the local frame of the jth aperture can be described by (x_j, y_j), and $T(x,y)$ represents now the transmission function corresponding to each individual region. Without loosing from generality, we can consider that the apertures are described by the same individual distribution function, and thus one can exchange summation with integration. Thus, Eq. (3.30) can be rewritten as:

$$A(p, s) = \int\int_{-\infty}^{+\infty} T(x, y)e^{-2i\pi(px+sy)}\,dx\,dy \times \sum_{j=1}^{N} e^{ipx_j}e^{isy_j}. \tag{3.31}$$

The first factor on the right hand side is the Fourier transform of the distribution function of each of the identical apertures. The second factor gives the Fourier transform of Dirac-delta distributions through $A_\delta = \sum_{j=1}^{N}(x - x_j)(y - y_j)$, and it takes into account the spatial distribution of the apertures inside the array. Therefore, Eq. (3.31) can be rewritten as:

$$A(p, s) = \mathscr{F}\{T(x, y)\}\mathscr{F}\{A_\delta\}, \tag{3.32}$$

which is known as the array theorem [5], and it states that the field distribution of Fraunhofer diffraction from an array of similarly oriented, identical apertures is the product of the field distribution of the Fraunhofer diffraction from any one of the apertures with the Fourier transform of the set of delta functions distributed in the same manner as the apertures in the array.

Then, the intensity distribution of the diffracted beam, can be written as:

$$I(p, s) \equiv |A(p, s)|^2 = |\mathscr{F}\{T(x, y)\}|^2|\mathscr{F}\{A_\delta\}|^2. \tag{3.33}$$

The first factor gives the scattering intensity of a single aperture, while the second one reveals the way in which they are distributed. In SAS, the components $\{p, s\}$ are generally denoted by $\{q_x, q_y\} \equiv \mathbf{q}$. These quantities are also known as the form factor $P(q)$ and, respectively, the structure factor $S(q)$. Thus formally, in terms of the module of the scattering wave vector q, the intensity becomes:

$$I(q) \propto P(q)S(q). \tag{3.34}$$

In addition, in a physical SAS experiment using x-rays or neutrons, the scattering power of the objects, as well as their different orientations shall be taken into account. As we shall see below, in the former case, this appears as a constant factor in the right hand side of Eq. (3.34), while in the later case, this is introduced through an ensemble averaging over all possible orientations.

3.5 Scattering Lengths, Scattering Length Densities and Contrast

In an atom each electron has the same scattering length and thus the scattering from an element is proportional to the number of its electrons Z, that is [1]:

$$b_x = b_0 Z, \tag{3.35}$$

The numerical value of b_0 for the Thomson scattering factor of one electron is obtained by considering that the scattered amplitude E_s of a plane wave from an electron can be written as $E_s = -E_0 e^2/(m_e c^2 r) \sin \alpha$. Here, E_0 is the amplitude of the incident wave, e and m_e are the charge, and respectively the mass of the electron, c is the light velocity, r is the distance from electron to the detector, and α is the angle between the scattered beam and the direction of electron displacement. In the case of SAXS, $\alpha = \pi/2$, and using Eq. (3.6), one obtains $b_0 = 0.282 \times 10^{-12}$ cm. Table 3.1 lists the scattering lengths b_x for several common elements. It can be clearly seen that b_x increases with the atomic number Z.

For a complex system, consisting from various types of atoms, the corresponding scattering length is the sum of the scattering lengths of the composing atoms, i.e. [1]:

$$b_{\text{tot}} = \sum_{i=1}^{n} b_i, \tag{3.36}$$

where n is the number of atoms. Note that according to Eq. (3.35), in an experiment the signal will be given mostly by the scattering from heavy atoms. As an application, by using Eq. (3.36) together with values of b_x from Table 3.1 for carbon and oxygen, the scattering length of toluene ($C_7 O_8$) becomes: $b_{\text{tot}} = (7 \times 1.69 + 8 \times 2.26) \times 10^{-12}$ cm $= 29.91 \times 10^{-12}$ cm. Similarly, for water ($H_2 O$) and heavy water ($D_2 O$) we can see that the x-ray scattering lengths coincide and have the value of $b_{\text{tot}} = 2.82 \times 10^{-12}$ cm.

In the case of SANS, each atom is characterized by a neutron scattering length, which determines the strength of neutron scattering, and represents the apparent "size" of the atom during scattering. It varies irregularly with isotopes, since it depends on the type of nucleus as well as on the total spin of nucleus-neutron system. The thermal neutron wavelength is much larger than the dimensions of nuclei, and this implies that the corresponding scattering length is isotropic and independent of the scattering angle.

Formally, neutron scattering lengths can be expressed as [1]:

$$b_n = \langle b \rangle \pm \Delta b, \tag{3.37}$$

where $\langle b \rangle$ is the coherent scattering length, i.e. the average scattering length over all isotopes and spin state populations, and Δb is the incoherent scattering length, i.e. the root mean square deviation of scattering length b_n from its average $\langle b \rangle$. The values of coherent neutron scattering lengths are obtained experimentally, and their values for some common elements are listed in Table 3.1. Besides irregular variation with atomic number, the main feature is that in the case of hydrogen (as well for some other elements not listed in Table 3.1) $\langle b \rangle$ has a negative value. This means that neutrons scattered from hydrogen have a 180° phase shift as compared to other atoms. For comparison, the neutron scattering length for toluene is $b_n = (7 \times 0.665 + 8 \times$

Table 3.1 X-ray scattering lengths (b_X), neutron scattering lengths (b_n) and neutron coherent scattering cross-sections (σ_{coh}) as a function of atomic number Z for several atoms

Atom	Z	b_x (10^{-12} cm)	b_n (10^{-12} cm)	σ_{coh} (10^{-24} cm^2)
H	1	0.282	−0.374	1.76
D	1	0.282	0.667	5.59
C	6	1.69	0.665	8.36
N	7	1.97	0.94	11.81
O	8	2.26	0.58	7.29
Na	11	3.10	0.36	4.52
Si	14	3.94	0.42	5.28
Ca	20	5.64	0.47	5.91
Fe	26	7.33	0.95	11.94
Co	27	7.61	0.28	3.52

0.580) $\times 10^{-12}$ cm = 22.74 $\times 10^{-12}$ cm, for water $b_n = -0.17 \times 10^{-12}$ cm, while for heavy water $b_n = 1.91 \times 10^{-12}$ cm.

In SANS, the difference in the scattering lengths of hydrogen and deuterium is used for the so-called deuterium labelling, and thus SANS has a considerable advantage over SAXS. Considering the high penetration power of neutrons and high interaction probability, mainly with light elements (see Table 3.1), SANS is a widely used technique for studying the structure of organic compounds.

Furthermore, for neutron scattering the coherent and incoherent scattering cross sections can be written as [3]:

$$\sigma_{coh} = 4\pi \langle b \rangle^2 \quad \text{and} \quad \sigma_{incoh} = 4\pi \langle \Delta b \rangle^2 . \tag{3.38}$$

The structure of the sample is determined only from the coherent scattering since it contains information concerning interference phenomena. Table 3.1 lists the coherent scattering cross sections for several common elements, obtained using Eq. (3.38). The incoherent component represents only the background, and which has to be subtracted from experimental data, unless the interest resides on the motion of single atoms and molecules. In this case are necessarily experiments performed in inelastic mode, involving changes in the energy of the incident beam. However, this topic is beyond the scope of this book.

In a homogeneous sample, i.e. where the scattering power is independent of the position of the scattering center, radiation is scattered only in the forward direction ($\theta = 0°$). Therefore, at $\theta \neq 0°$, a non-zero signal is observed only when there exists a variation of the scattering power throughout the sample. When the volume of such inhomogeneities is of the order of radiation wavelength, the corresponding SAS signal will be observable at low and intermediate values of the scattering vector $q \lesssim 1$ Å^{-1}. In order to quantify the scattering power of a volume v, one introduces the quantity:

$$\rho = \frac{1}{v} \sum_{i=1}^{n} b_i, \tag{3.39}$$

called the scattering length density (SLD). Here, n represents the number of atoms in a molecule, b_i is the coherent scattering length given by Eq. (3.37) in the case of neutron scattering, and the scattering length given by Eq. (3.36) in the case of x-rays scattering. The volume v is usually considered that of a single molecule, and thus:

$$v = \frac{1}{N_A \eta} \sum_{i=1}^{n} M_i, \tag{3.40}$$

where $N_A = 6.022 \times 10^{26}$ mol^{-1} is the Avogadro's number, η is the density, and M_i is the atomic mass of the ith atom. In the case of toluene, the density is 0.87 g/cm^3, $n = 2$, $\sum_{i=1}^{2} M_i = 7 \times 12.01 + 8 \times 1.0075 = 92.13$ g/mol, and thus the molecular volume is $v = 177 \times 10^{-24}$ cm^3. Therefore, using Eq. (3.39) one finds that neutron SLD of toluene is $\rho_n = 0.94 \times 10^{10}$ cm^{-2}. Also, using data from Table 3.1, one finds that for x-rays the SLD is $\rho_x = 8.1 \times 10^{10}$ cm^{-2} which is higher with about one order of magnitude than ρ_n. By performing similar calculations for water and heavy water, one finds that the corresponding neutron SLD are -0.56×10^{10} cm^{-2}, and respectively 6.39×10^{10} cm^{-2}. From another hand, the neutron scattering lengths densities of several solids such as carbon, aluminum oxide, silicone dioxide and many others are between those of water and heavy water. Thus, in a three-phase system, one can match the SLD of a given phase by changing the volume concentration of water + heavy water mixture, and thus basically dealing with a two-phase system. Experimentally, in a two-phase system, the scattering intensity is proportional to the square of the scattering contrast, i.e.:

$$I(q) \propto (\Delta \rho)^2, \tag{3.41}$$

where $\Delta \rho \equiv \rho_m - \rho_p$ is the difference between the SLD of one phase (i.e. a matrix) ρ_m and the SLD of the second phase (i.e. particles inside the matrix) ρ_p. It is clear from Eq. (3.41) that at matching point $\rho_m = \rho_p$ and thus $I(q) \propto 0$.

Note that when a beam irradiates a sample, a fraction of it is also absorbed and converted into other forms of energy. Therefore, the emergent beam is attenuated due to both scattering and absorption. Neutrons are trapped by nuclei, which become excited, and then they decay into a ground state by emitting γ-rays or α-particles by nuclear fission. X-rays excite the atoms which further reach the ground state by emitting electrons from their inner shells with energies close to those of the absorbed photon. However, this secondary radiation is not recorded in a SAS experiment [3]. As in the case of scattering, absorption is described in terms of the number of particles absorbed per second by a nucleus/atom divided to the total flux of the incident beam, which is known as the absorption cross section. Transmission experiments give information about the absorption edge, absolute absorption and on the average density [3, 6].

3.6 Obtaining Structural Parameters from SAS Curves

Without loosing from generality, let's consider that the sample is defined in terms of a density distribution of electrons or nuclei, instead of a discrete distribution. Then, the Fourier transform of this distribution gives the scattering amplitude of the whole irradiated volume V, and thus one can write (see Eq. (3.24)) [7]:

$$A(q) = \int_V \rho(r)e^{-iq\cdot r}dV. \tag{3.42}$$

where $\rho(r)$ is the scattering length density distribution of the sample. In the case of x-rays scattering, $\rho(r)$ is equal to the electron density distribution multiplied by the x-ray scattering length of one electron, while for neutron scattering, it is given by Eq. (3.39).

Then the product between the scattering amplitude $A(q)$ and its complex conjugate $A(q)^*$ gives the scattering intensity $I(q)$, expressed as [7]:

$$I(q) \equiv A(q)A(q)^* = \int \rho(u')e^{-iq\cdot u'}du' \int \rho(u)e^{-iq\cdot u}du. \tag{3.43}$$

In writing Eq. (3.43) it is assumed that multiple scattering is neglected, and thus we work in the framework provided by the kinematic theory. This is opposed to the situation encountered in crystals, where the beam can be diffracted many times before it leaves the crystal, and which requires to consider a dynamic theory of scattering. However, the later case is beyond the topic of this book.

By performing the change of variable $r = u' - u$, we obtain:

$$I(q) = \int [\rho(u)\rho(u+r)du]\, e^{-iq\cdot r} = \int \Gamma(r)e^{-iqr}dr, \tag{3.44}$$

where $\Gamma(r)$ is the autocorrelation function of $\rho(r)$, that is the convolution square of the density fluctuations, and is given by:

$$\Gamma(r) = \int \rho(u)\rho(u+r)du. \tag{3.45}$$

This quantity specifies how the densities $\rho(u)$ and $\rho(u')$, separated by the distance r, are correlated to each other on the average. Then, the spherically averaged autocorrelation function becomes:

$$\Gamma(r) = \left\langle \int \rho(u)\rho(u+r)du \right\rangle. \tag{3.46}$$

The correlation (characteristic) function $\gamma(r)$ is the normalized version of the autocorrelation function, and is given by [1, 7]:

$$\gamma(r) = \frac{\Gamma(r)}{\Gamma(0)}, \tag{3.47}$$

so that, $\gamma(0) = 1$, with $\Gamma(0) = V\rho^2$.

As seen before, in a SAS experiment, one can generally consider a two-phase model, in which scatterers/particles are embedded in a matrix/solvent. Then, the SLD which appears in Eq. (3.42) shall be replaced by the excess SLD. When particle's inhomogeneities are much larger than those of the solvent, the SLD appearing in Eq. (3.42) shall be replaced by the excess SLD, given by $\Delta\rho(r) = \rho(r) - \rho_0$, where $\rho(r)$ is the SLD of the particles and ρ_0 is the SLD of the solvent/matrix. Therefore, $\Delta\rho(r)$ can take both positive and negative values. In the case of highly diluted systems, a mean particle density can be considered, and the above difference gives the contrast $\Delta\rho$ (see Eq. (3.41)), which is a very important parameter in structural investigations.

Taking into account that the average values $\langle \cdots \rangle$ of the exponential function in Eq. (3.43) are given by [1]:

$$\left\langle e^{-iq\cdot r} \right\rangle = \frac{\sin qr}{qr}, \tag{3.48}$$

then it can be rewritten as [7, 8]:

$$I(q) = 4\pi V \int_0^\infty r^2 \gamma(r) \frac{\sin qr}{qr} dq \tag{3.49}$$

or, in terms of the pair distance distribution function $p(r) = r^2 V\gamma(r)$, as:

$$I(q) = 4\pi \int_0^\infty p(r) \frac{\sin qr}{qr} dr. \tag{3.50}$$

The physical meaning of $p(r)$ can be worked out more clearly if we subdivide the particle into a large number of smaller volume elements. Then, $p(r)$ is related to the number of lines with lengths between r and $r + dr$, and which is obtained by taking into account arbitrarily combinations of distinct volume elements inside the particle. An important property of is that $p(r) = 0$ at $r = 0$ and for $r > D$, where D is the maximum distance in the particle. As a consequence, in the case of homogeneous particles $p(r)$ represents a histogram of distances. However, it does not contain any information about the orientations of these lines, since an averaging over orientations is performed. For inhomogeneous particles, we may have negative contributions to $p(r)$, since we have to weight each line with the product of the difference between density distribution and the differential volume element. A minimum in $p(r)$ can be given either by a small number of distances or by addition of positive and negative contributions.

By using the previous relation between the pair distance distribution and the correlation function, the inverse Fourier transform of $p(r)$ and $\gamma(r)$ become [8]:

$$p(r) = \frac{1}{2\pi^2} \int_0^\infty I(q)qr \sin(qr)dq, \tag{3.51}$$

and, respectively:

$$V\gamma(r) = \frac{1}{2\pi^2} \int_0^\infty I(q)q^2 \frac{\sin qr}{qr}dq. \tag{3.52}$$

The correlation function $\gamma(r)$ also depends on the particle's geometry, and expresses the set of distances joining the volume elements within the particle, as well as the distribution of the inhomogeneities within the particle. Since, both $p(r)$ and $\gamma(r)$ are connected to the scattering intensity, they are very important for the estimation of the structure from the corresponding scattering intensities.

Below we present the main parameters which are extracted either from the scattering intensity $I(q)$ or directly prom $p(r)$ function. For this purpose, we consider first a simple case of homogeneous particles with the same size, shape and internal structure at infinite dilution, that is $S(q) = 1$ in Eq. (3.34). This situation is very often encountered in biological macromolecules in solution. Thus, the density can be written as:

$$\rho(r) = \begin{cases} \rho, & \text{if } r \in V \\ 0, & \text{otherwise.} \end{cases} \tag{3.53}$$

One of the most important structural parameter is the radius of gyration R_g, also known as the second moment of inertia. It gives the mass distribution of a particle around its center of gravity. In the case of SAS from macromolecules, R_g can be used to characterize the conformational changes.

We saw in Eq. (3.50) that the scattering intensity can be expressed through a Fourier transform of the pair distance distribution function. The sine term in this equation can be approximated by a Taylor series expansion, such as:

$$\sin(qr) \simeq \sin a + \frac{\cos a}{1!}(qr - a) - \frac{\sin a}{2!}(qr - a)^2 - \frac{\cos a}{3!}(qr - a)^3 + \cdots. \tag{3.54}$$

Since experimental data are in close proximity of the zero angle, we set $a = 0$ in Eq. (3.54) and write:

$$\sin(qr) \simeq qr - \frac{1}{3!}(qr)^3 - \frac{1}{5!}(qr)^5 + \cdots. \tag{3.55}$$

In addition, since q is very small, it is enough to keep only the first two terms in Eq. (3.55). Thus, we can rewrite Eq. (3.50) as:

$$\begin{aligned}
I(q) &\simeq 4\pi \int_0^\infty p(r)\frac{1}{qr}\left(qr - \frac{1}{3!}(qr)^3 - \cdots\right)dr \\
&\simeq 4\pi \int_0^\infty p(r)dr - 4\pi\frac{1}{3!}\int_0^\infty p(r)(qr)^2 dr + \cdots
\end{aligned}$$

$$\simeq 4\pi \int_0^\infty p(r)\mathrm{d}r - 4\pi \frac{q^2}{3!} \frac{\int_0^\infty p(r)\mathrm{d}r}{\int_0^\infty p(r)\mathrm{d}r} \int_0^\infty p(r)r^2 \mathrm{d}r + \cdots$$

$$\simeq 4\pi \int_0^\infty p(r)\mathrm{d}r \left(1 - \frac{q^2}{3} \frac{1}{2} \frac{\int_0^\infty p(r)r^2 \mathrm{d}r}{\int_0^\infty p(r)\mathrm{d}r} + \cdots \right)$$

$$\simeq I(0) \left(1 - q^2 \frac{R_g^2}{3} + \cdots \right), \tag{3.56}$$

where, we have:

$$I(0) = 4\pi \int_0^\infty p(r)\mathrm{d}r, \tag{3.57}$$

and

$$R_g^2 = \frac{\int_0^\infty p(r)r^2 \mathrm{d}r}{2 \int_0^\infty p(r)\mathrm{d}r}. \tag{3.58}$$

We can see that in the last equality of Eq. (3.56) the parenthesis contains the expansion terms of the Taylor series of the function e^x, with $x = -q^2 R_g^2/3$. Therefore, Eq. (3.56) can be written more compact as:

$$I(q) \simeq I(0)e^{-\frac{q^2 R_g^2}{3}}, \tag{3.59}$$

which is called the Guinier approximation, and which holds at small values of the wavevector q. Taking the logarithm of both sides, we obtain:

$$\ln I(q) = \ln I(0) - \frac{R_g^2}{3} q^2. \tag{3.60}$$

Therefore, by performing a linear fit of Eq. (3.60) in a Guinier plot, that is a plot of $I(q)$ *versus* q^2, both the intensity at zero angle $I(0)$ and the radius of gyration R_g can be determined from the intercept, and respectively from the slope of the fitting function. As a rule of thumb, the end of the fitting (Guinier) region shall be chosen in such a way that $q_{max}R_q \lesssim 1.3$. We shall see in Sect. 3.10 an application of Guinier analysis to two macromolecular systems: lysozyme an glucose isomerase.

The Guinier analysis is often the first step in basic SAS data processing. Usually it is complemented by a molecular weight analysis together with a Kratky plot in order to asses other structural features, which will be discussed below. Depending on the type of the sample and on the sought parameters, more advanced analysis can be performed, such as: size-exclusion chromatography SAS processing, wide angle scattering—SAS data merging, pair-distance distribution analysis or 3D reconstruction with bead (dummy atom) models [9].

A molecular weight analysis can be performed in several ways, depending mainly on the investigated system. The most common ones are: referencing $I(0)$ to that of a known standard [10], using the correlation volume [11], using the adjusted Porod volume [12] or by using the value of $I(0)$ on an absolute scale. For illustration purposes one considers in Sect. 3.10 the first method, where $I(0)$ is proportional to the molecular weight, concentration and contrast of a macromolecule in solution. If a reference sample of known molecular weight $M\,W_{st}$ and concentration c_{st} is measured, then it can be used to calibrate the molecular weight $M\,W_m$ of any other profile with known concentration c_m as [10]:

$$M\,W_m = \frac{(I(0)_m/c_m)\,M\,W_{st}}{I(0)_{st}/c_{st}}. \tag{3.61}$$

A Kratky analysis involves representing SAS data in a plot of $q^2 I(q)$ versus q, and can provide information about the degree of unfolding in a macromolecular solution. In a Kratky plot, unfolded, i.e. highly flexible molecules, have a plateau at high q, that is a region where $q^2 I(q)$ is constant, while globular molecules have a bell-shaped peak. A system consisting of partially unfolded molecules can have both the plateau and the bell-shaped region (See 3.10).

Furthermore, by using Eqs. (3.57) and (3.1), it can be easily shown that the volume of a homogeneous particle is given by:

$$V = 2\pi^2 \frac{I(0)}{Q}, \tag{3.62}$$

where $Q = \int_0^\infty I(q)q^2 dq$ is called the Porod invariant. The surface S of the particle is determined from the high q-region of the scattering curve, through [1]:

$$S = \frac{\lim_{\to\infty} \left(I(q)q^4\right)}{2\pi(\Delta\rho)^2}, \tag{3.63}$$

where $\Delta\rho$ is the scattering contrast. Therefore, the last two quantities can be used to obtain the specific surface S/V. Similarly, one can find the area of the cross-section and the thickness of the lamellar particles.

An important parameter which can be obtained from the correlation function is the correlation length l_c given by [1]:

$$l_c = \frac{\pi}{Q} \int_0^\infty I(q)q dq, \tag{3.64}$$

which represent the mean width of $\gamma(r)$.

Depending on the specific shape of the particle, other structural parameters can be obtained from SAS curves, such as: the persistence length, mas per unit length of rod-like particles or mass per unit area of flat particles. As we already have seen above, some properties are discussed in terms of the scattering curve while others in terms of the pair distance distribution function or correlation function. Generally, the shape and structure of a particle are better understood using real space functions, while other characteristics, such as symmetry, are better represented in the reciprocal space.

In the case of composite structures consisting of aggregates with a large number of identical subunits, the overall structure is obtained from the beginning of the scattering curve, while the influence of the subunits is seen at larger values of the scattering vector. For particles with internal inhomogeneities it is not possible to recover a unique shape from the scattering curve. However, additional structural properties about inhomogeneous particles is obtained by contrast variation [13].

A great number of other samples have a high concentration of particles and require a different approach for analyzing the corresponding scattering data. Mainly, interference terms shall be taken into account, that is $S(q) \neq 1$ in Eq. (3.34), and thus information on the spatial arrangements of particles can be obtained. For this purpose, one can consider an isotropic system of monodisperse spheres, where each sphere has the same surroundings, and is characterized by radial interparticle distribution function $P(r)$. Thus, the main question is to find the probability that one particle, situated at distance r apart from another (central) particle, will be found in the volume element dV.

If the system consists of N particles, each of volume v, then $P(r)$ will describe deviations of this probability from the mean value $(N/v)dV$. Considering that the particle's size is D, and the SLD satisfy Eq. (3.53), then the radial distribution function is related to the scattering through [14]:

$$I(q) = NI_0(q) \left(1 + \frac{N}{V} \int_0^\infty 4\pi r^2 \left(P(r) - 1 \right) \frac{\sin(qr)}{qr} dr \right), \qquad (3.65)$$

where $I_0(q)$ is the scattering intensity of a single particle. Therefore, if $I_0(q)$ is known, $P(r)$ can be obtained from experimental data by dividing Eq. (3.65) to $I_0(q)$ and performing a Fourier transform. Thus, $P(r)$ will contain all the interparticle interferences, having the following properties: $P(r) = 0$ for $r < D$, and $P(r) = 1$ for $r \gg D$. The general behaviour of $P(r)$ is similar to an oscillating function with decreasing amplitudes, and reaching $P(r) = 1$ at high values of r. From another hand, if we are interested to find the scattering intensity $I_0(q)$ in the presence of interference effects, then the later ones have to be eliminated in Eq. (3.65). Experimentally, this is performed by measuring the scattering intensity at different concentrations c, and then to extrapolate $I(q)/c$ curves to zero concentration. However, this procedure is not always possible, since the tendency of agglomeration may be present at very high dilution [14].

3.7   Influence of Experimental Factors and Orientational Averaging on the Shape of SAS Curve

Besides the shape and size of scattering objects, the main factors affecting the behaviour of a scattering curve are: averaging procedures, object polydispersity and instrumental resolution. Averaging is related to the orientations and shapes of the objects inside the sample, polydispersity to the sizes of the particles, and the resolution is related to deviations of the beam from the ideal case and respectively to limitations of the scattering device.

3.7.1 Orientational Averaging

When a macroscopic volume is investigated, as is in the case of SAS experiments, the investigated contains a large number of particles with random orientations, and thus the scattering intensity shall be averaged over all these possible orientations. If the probability of any orientation is the same, then it can be calculated by averaging over all directions n of the momentum transfer $q = qn$, that is, by integrating over the solid angle in the spherical coordinates [15]:

$$\langle f(q_x, q_y, q_z) \rangle = \frac{1}{4\pi} \int_0^\pi d\theta \sin\theta \int_0^{2\pi} d\phi f(q, \theta, \phi), \tag{3.66}$$

where, in spherical coordinates $q_x = q \cos\phi \sin\theta$, $q_y = q \sin\phi \sin\theta$, $q_z = q \cos\theta$.

If the investigated object is allowed to rotate only in a two-dimensional space, then the averaging is performed [16]

$$\langle f(q_x, q_y) \rangle = \frac{1}{2\pi} \int_0^{2\pi} f(q, \phi) \, d\phi, \tag{3.67}$$

where $q_x = q \cos\phi$ and $q_y = q \sin\phi$. Depending on the symmetry of the object, Eqs. (3.66) and (3.67) can be further simplified. Therefore, by taking into account the concentration c of the objects inside the sample, the SLD (see Eq. (3.41)), and the orientational average provided by by Eqs. (3.66) and (3.67) we obtain a general expression for the scattering intensity of monodisperse objects, such as:

$$I(q) = c|\Delta\rho|^2 V^2 \langle |F(q)|\rangle^2, \tag{3.68}$$

where V is the volume if each object. We shall see in the next sections, that for fractal systems, the averaging procedures lead to a power-law decay of the scattering intensity, with the exponent related to the fractal dimension of the fractal.

3.7.2 Polydispersity

In a physical system, almost always there exists a certain degree of polydispersity. As for the monodisperse case, we can consider two cases: without and with interparticle interference. Let's consider that the sizes follow a distribution function $D_N(l)$, which is defined in such a way that $D_N(l)dl$ gives the probability of finding an object whose size falls within $(l, l + dl)$. We consider here log-normal distribution, defined by:

$$D_N(l) = \frac{1}{\sigma l(2\pi)^{1/2}} e^{-\frac{\left(\ln(l/l_0)+\sigma^2/2\right)^2}{2\sigma^2}}, \tag{3.69}$$

but any other "broad" distribution can be considered. Here, $\sigma = \left(\ln(1+\sigma_r^2)\right)^2$ is the variance, $l_0 = \langle l \rangle_D$ is the mean value of the length, $\sigma_r \equiv \left(\langle l^2 \rangle_D - l_0^2\right)^{1/2}/l_0$ is the *relative* variance, and $\langle \cdots \rangle_D = \int_0^\infty \cdots D_N(l)dl$. Thus, the polydisperse scattering intensity is obtained by averaging Eq. (3.68) over the distribution function (3.69) [15]

$$I(q) = c|\Delta\rho|^2 \int_0^\infty \left\langle |F(q)|^2 \right\rangle V(l)^2 D_N(l)dl, \tag{3.70}$$

$F(q) \equiv A(q)/V$ (see Eq. (3.42)) is the normalized scattering form factor, and the symbol $\langle \cdots \rangle$ stands for ensemble averaging over all orientations (see below). The general effect of polydispersity is to smooth the oscillations present in the scattering intensity, i.e. the stronger the polydispersity the smoother the curve [17]. However, when interparticle interferences are present, the situation is more complicated, and no general method exists to describe such cases.

Also, experimentally, the beam has a finite size, it contains radiation of different wavelengths, it is not perfectly collimated, and the detector has a finite size. All these effects lead to a smearing of SAS data. The inverse process, i.e. desmearing of experimental data ca be achieved by using advanced mathematical procedures, such as indirect transformation methods [18, 19].

3.8 Derivation of Form Factors of Basic Euclidean Shapes

For most of the particles having a well-defined geometrical shape it is possible to calculate analytically the corresponding scattering intensity. As we already discussed in Sect. 3.6, the intensity is obtained from the product of the scattering amplitude with its complex conjugate (see Eq. (3.43)). In turn, the scattering amplitude is related by a Fourier transform to the SLD $\rho(r)$. In the following we shall illustrate the process of calculating the scattering intensity, as well as the pair distance distribution and correlation functions for few basic geometrical shapes, and discuss their properties.

3.8.1 Scattering from Three-Dimensional Structures: Spheres

We start with a sphere of radius R, since this is a very common shape and due to its symmetry, it does not require any orientational average. Thus, if the sphere has a uniform density ρ_0, the corresponding SLD can be written as:

$$\rho(r) = \begin{cases} \rho_0, & r \leq R \\ 0, & r > R. \end{cases} \tag{3.71}$$

In order to evaluate the scattering amplitude with SLD given by Eq. (3.71), we rewrite Eq. (3.42) in Cartesian coordinates x, y and z in real space, such as:

$$A(q_x, q_y, q_z) = \int_{\infty}^{\infty} \int_{-\infty}^{\infty} \int_{-\infty}^{\infty} \rho(x, y, z)e^{-(q_x x + q_y y + q_z z)}dxdydz, \tag{3.72}$$

where $r = (x, y, z)$, q_x, q_y, q_z are the components of vector q in reciprocal space, that is $q = \{q_x, q_y, q_z\}$, and $dr = dxdydz$ is the volume element in 3D space. Due to the symmetry of the object, the calculations can be greatly simplified if we work in spherical polar coordinates, that is we express the position of a point in the sphere in terms of three variables r, θ and ϕ, satisfying the conditions: $0 \leq r < \infty$, $0 \leq \theta \leq \pi$ and respectively $0 \leq \phi \leq 2\pi$. Therefore, the volume element is given by $dr = r^2 \sin\theta dr d\theta d\phi$, and Eq. (3.72) becomes:

$$A(q) = \int_0^{2\pi} \int_0^{\pi} \int_0^{\infty} \rho(r, \theta, \phi)e^{-iq\cdot r}r^2 \sin\theta dr d\theta d\phi. \tag{3.73}$$

From another hand, Eq. (3.71) clearly shows that $\rho(r, \theta, \phi)$ is constant on the whole sphere, that is, it does not depend on the variables r, θ and ϕ, and therefore we can denote $\rho_0 = \rho(r, \theta, \phi)$. Thus, we can calculate the integral in Eq. (3.73) by choosing the polar axis to coincide with the direction of the vector q, that is letting $q \cdot r = qr \cos\theta$. Furthermore, by performing the change of variable $u = \cos\theta$, we have $u = 1$ for $\theta = 0$, $u = -1$ for $\theta = \pi$, and $du = -\sin\theta d\theta$. Then, Eq. (3.73) can be rewritten as:

$$A(q) = \rho_0 \int_0^{2\pi} \int_{-1}^{1} \int_0^{\infty} e^{-iqru}r^2 dr du d\phi. \tag{3.74}$$

After little algebra, one can write Eq. (3.74) as:

$$A(q) = 2\pi\rho_0 \int_0^{\infty} r^2 \frac{e^{iqr} - e^{-iqr}}{iqr}dr = 4\pi\rho_0 \int_0^{\infty} r^2 \frac{\sin(qr)}{qr}dr. \tag{3.75}$$

Taking into account that SLD is constant for $r \leq R$, we rewrite Eq. (3.75) such as:

$$A(q) = \frac{\rho_0}{q} \int_0^{R} 4\pi \sin(qr)dr. \tag{3.76}$$

Performing integration by parts, the last equation yields:

$$A(q) = v\rho_0 \frac{3\left(\sin(qR) - qR\cos(qR)\right)}{(qR)^3}, \tag{3.77}$$

Table 3.2 Radius of gyration R_g and normalized scattering intensities $I(q)/I(0)$ for some basic shapes used throughout the book. Here, $J_1(q)$ is the spherical Bessel function

Object	R_g^2	$I(q)/I(0)$	Refs.	Notes
Sphere of radius R	$(3/5)\,R^2$	$9\left(\frac{\sin(qR)-qR\cos(qR)}{(qR)^3}\right)^2$	[1]	Eq. (3.78)
Parallelepiped of edges a, b, c	$\frac{a^2+b^2+c^2}{12}$	$\left(\frac{q_x a/2}{q_x a/2}\right)^2\left(\frac{q_y a/2}{q_y a/2}\right)^2\left(\frac{q_z a/2}{q_z a/2}\right)^2$	[1]	–
Disk of radius R	$\frac{R^2}{12}$	$4\frac{J_1^2(q)}{q^2}$	[14]	–
Square of edge a	$\frac{a^2}{2}$	$\left(\frac{q_x a/2}{q_x a/2}\right)^2$	[1]	–
Triangle of height h	$\frac{h^2}{18}$	$\left(2e^{-\alpha}\,\frac{\beta e^{i\alpha}-\beta\cos\beta-i\alpha\sin\beta}{\beta(\beta^2-\alpha^2)}\right)^2$	[21]	$\alpha = hq_x$ $\beta = hq_x^2/2$

where $v = (4/3)\pi R^3$ is the volume of the sphere. Finally, according to Eq. (3.43), the scattering intensity from a sphere becomes:

$$I(q) = v^2 \rho_0^2 \frac{9\,(\sin(qR) - qR\cos(qR))^2}{(qR)^6}. \tag{3.78}$$

Following a similar approach, the scattering amplitudes and intensities for other simple shapes can be determined analytically. Table 3.2 at the end of this section lists the most common expressions for the scattering intensities, together with the corresponding radii of gyration.

Figure 3.2 shows the behaviour of scattering intensity for $R = 10$ nm on a double logarithmic scale. The main feature is the presence of a Guinier region, followed by a Porod one. The end of the Guinier region is marked by a vertical line, and from its position an estimation of the overall size of the sphere can be obtained. The Porod region is characterized by a nearly periodic behaviour where the minima are given by the condition $\sin(qr) - qr\cos(qr) = 0$, that is $qR \simeq (2k + 1)\pi/2$, with $k \in \mathbb{N}^*$. Note, that the presence of minima in the scattering curve is characteristic to object with higher symmetry, that is the minima are sensibly smoothed in the case of cubes, due to lower symmetry. Also, a change in the radius leads to a horizontal shift of the whole scattering curve either to left (for $R > 10$ nm) or to the right (for $R < 10$ nm).

The same figure shows the effect of polydispersity given by Eq. (3.70), where the sizes of the spheres are distributed according to Eq. (3.69). It is clear that by increasing the relative variance σ_r, the scattering curves become smoother, and resemble more closely those obtained in a SAS experiment. The polydisperse curve reveals also that in the Porod region, the decay of the intensity is proportional to q^{-4}. Similar calculations for 2D objects show that in the Porod region the intensity decays proportional to q^{-3} while for 1D objects, it is proportional to q^{-1}.

As for the scattering intensity, analytic expressions for the correlations and pddf functions are available for basic geometric shapes. In the case of more complex

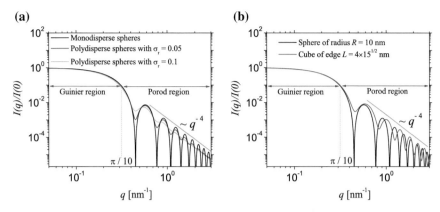

Fig. 3.2 SAS from: **a** Monodisperse and polydisperse systems of spheres (Eq. (3.78)) at various relative variances σ_r of the log-normal distribution given by Eq. (3.69). Lower (black) curve: $\sigma_r = 0$; Middle (blue) curve: $\sigma_r = 0.05$; Upper (green) curve: $\sigma_r = 0.1$ **b** Sphere (lower-black curve) and cube (upper-red) of the same radius of gyration. The normalized scattering amplitude of the cube is given in Table 3.2. The average over orientations for the cube has been performed according to Eq. (3.66). Vertical (magenta) dotted-lines delimitates the Guinier and Porod regions

shapes, correlation and pddf functions can be determined by various numerical methods such as Monte Carlo simulations. As an example, for the same sphere of radius R we have [8]:

$$\gamma(r) = 1 + \frac{3}{4}\frac{r}{R} + \frac{1}{16}\frac{r^3}{R^3}, \tag{3.79}$$

while for a cube of edge length L, we have [20]:

$$\gamma(r) = 1 - \frac{3}{2}\frac{r}{L} + \frac{2}{\pi}\frac{r^2}{L^2} - \frac{1}{4\pi}\frac{r^3}{L^3}, \tag{3.80}$$

for $0 \leq r \leq L$.

Figure 3.3 shows the correlation and pddf functions of a sphere of radius $R = 10$ nm and of a cube of a the same radius of gyration. The correlation functions are very similar, excepting that for a cube it decays slightly more concave. The pddf of the sphere has a maximum near $r = R$, and drops to zero at $r = 2R = 20$ nm, while for a cube, although the pddf is very similar to that of the sphere and the location of the maximum is the same, the pddf attains its zero at about 22 nm. Generally, at a constant R_g and $I(0)$, any deviation from the spherically symmetry, is seen as a shift of the maximum to smaller values of r, and the value of the maximum distance in the particle increases. For object with even lower symmetry than that of a cube, the corresponding pddf will be significantly different than that of a sphere, and the interpretation of scattering functions in real space reveal more easily the structural features of the particle.

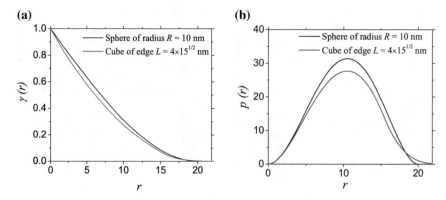

Fig. 3.3 **a** Correlations functions $\gamma(r)$ of a sphere (Eq. (3.79); upper curve) and of a cube (Eq. (3.80); lower curve) with the same radius of gyration. **b** The corresponding pair distribution functions of the sphere (upper curve) and of the cube (lower curve)

The scattering intensities for 2D objects is obtained from Eq. (3.43) but using the averaging provided by Eq. (3.67). The corresponding expressions are listed in Table 3.2. The corresponding correlation and pddf functions can be calculated from the Fourier transform of the corresponding intensities. For a triangle, we consider first an isosceles one of height $h = a\sqrt{3}/2$, where a is the base length a. Thus, the corresponding area is $S = ah/2$. Then we consider a Cartesian coordinate system with the x-axis parallel to the base, and the opposite vertex coincide with the origin. The corresponding and the normalized scattering amplitude can be written as [21]:

$$F^{T(q)} = \frac{1}{S} \int_0^a dy \int_{-ya/(2h)}^{ya/(2h)} e^{-i(q_x x + q_y y)} dx, \qquad (3.81)$$

where $q = \{q_x, q_y\}$. Performing the calculations in Eq. (3.81) one obtains an analytic expression of scattering intensity for an isosceles triangle. However, by choosing all angles to be equal, an equilateral triangle is obtained and the corresponding intensity is listed in Table 3.2.

3.9 Numerical Methods for Calculating Scattering Intensities

Although there exists a large number of simple shapes for which analytic expressions of the scattering amplitude are available, many of the physical samples are characterized by the presence of complex structures, such as aggregates, hierarchical structures or various types of fractals, and for which analytic expressions are generally not available. Also, implementing a similar procedure as in previous section in calculating the scattering intensity can lead to very complex integrals, which can be computed only numerically. In order to handle calculations of scattering intensities from such complex systems, one usually resorts to other approaches, which involves various approximations, but still reveal the main features of the scattering curve.

In the following subsections we present few methods, and we will use some of them in later sections to calculate the scattering intensity of fractals generated by various methods and to describe structural parameters from experimental data.

3.9.1 Debye Formula

A first method involves discretization of the object under investigation into smaller parts which are approximated by basic shapes (usually spheres). Then, the scattering intensity is computed by taking into account their shape and relative position within the object. This is handled using the Debye formula [22]. It is clear that the finer the discretization, the the better the approximation. A high degree of discretization leads to a very high number of points which have to be taken into account. Debye formula has the disadvantage that it has a high computational cost, i.e time increases with $\mathcal{O}(n^2)$. However, various optimization exists [23], including implementation on GPU hardware [24].

For development of theoretical models one can use chaos game representation (CGR) and cellular automata (CA) to generate positions of the N scattering basic units [25]. For experimental data with known atomic positions, such as those stored in the protein data bank [26], the CRYSOL program is usually used to calculate the scattering intensity [27].

The Debye formula reads as [22]:

$$I(q) = NI_0(q) + 2F_0(q)^2 \sum_{i=1}^{N-1} \sum_{j=i+1}^{N} \frac{\sin qr_{ij}}{qr_{ij}}, \qquad (3.82)$$

where $I_s(q)$ is the intensity scattered by each unit, and r_{ij} is the distance between units i and j. When the number of units exceeds few thousands, the computation of the term $\sin(qr_{ij})/(qr_{ij})$ is very time consuming, and thus it is handled via a pair-distance histogram $g(r)$, with a bin-width commensurate with the experimental resolution. Thus Eq. (3.82) becomes [23]

$$I(q) = NI_0(q) + 2F_0^2(q) \sum_{i=1}^{N_{bins}} g(r_i) \frac{\sin qr_i}{qr_i}, \qquad (3.83)$$

where $g(r_i)$ is the pair-distance histogram at pair distance r_i. For determining fractal properties we can neglect the form factor, and consider $I_0(q) = F_0^2(q) = 1$. Thus, Eq. (3.83) gives the structure factor:

$$I(q) \equiv S(q) = N + 2 \sum_{i=1}^{N_{bins}} g(r_i) \frac{\sin qr_i}{qr_i}. \qquad (3.84)$$

3.9.2 Monte Carlo Simulations

Another popular method involves generation of random points within the volume of the object, calculating the corresponding pair distance distribution function, and then the scattering intensity, according to Eq. (3.51) [28].

The first step consists in a stochastic generation of the pair distance distribution function $p(r)$, which has the advantage that is the output of several other methods, such as the Indirect Fourier transform, developed by Glatter. Then, the scattering intensity is calculated from Eq. (3.43), where $p(r)$ is normalized such that $\int p(r)\mathrm{d}r = 1$, so that the radius of gyration, given by Eq. (3.58), becomes:

$$R_g^2 = \frac{1}{2}\int_0^D r^2 p(r)\mathrm{r}, \qquad (3.85)$$

where D is the maximum distance between two arbitrarily points within the particle. Basically, $p(r)$ is obtained by partitioning the interval $0 \leq r \leq D$ into $M + 1$ histogram bins, which are indexed from 0 to M. In the second step, two points are chosen at random from inside a rectangular box which circumscribe the geometric object under investigation. In the third step, if both points belong to the object, the distance d between them is counted and added to the histogram. Otherwise, the points are discarded, d is not counted, and a new pair of points is selected. Finally, the procedure is repeated until the desired number of distances is obtained. Practically, a number of $10^7 \div 10^8$ distances are generally enough to obtain very good statistics.

3.9.3 Spherical Harmonics and Multipole Expansions

In this method one expresses SLD $\rho(\mathbf{r})$ as a series of spherical harmonics, which is parametrized according to [29]:

$$\rho(\mathbf{r}) = \sum_{r=0}^{\infty}\sum_{m=-l}^{l} \rho_{lm}(r)Y_{lm}(\omega), \qquad (3.86)$$

where $Y_{lm}(\omega)$ are spherical harmonics, and $\rho_{lm}(r)$ are the multipole coefficients given by:

$$\rho_{lm}(r) = \int_\omega Y_{lm}^*(\omega)\rho(\mathbf{r})\mathrm{d}\omega, \qquad (3.87)$$

with $Y_{lm}^*(\omega) = (-1)^m Y_{lm}(\omega)$. Then, the scattering amplitude is given by:

$$A(q) = \sum_{r=0}^{\infty}\sum_{m=-l}^{l} A_{lm}(q)Y_{lm}(\Omega), \qquad (3.88)$$

and the scattering intensity becomes [29]:

$$I(q) = 2\pi^2 \sum_{r=0}^{\infty} \sum_{m=-l}^{l} |A_{lm}(q)|^2, \tag{3.89}$$

where the coefficients A_{lm} are given explicitly by [1]:

$$A_{lm}(q) = (iq)^l \left(\frac{2}{\pi}\right)^{1/2} \sum_{p=0}^{\infty} \frac{(-1)^p \, r^{l+2p+2}}{2^p p! [2(l+p)+1]!!} q^{2p}. \tag{3.90}$$

Thus, the scattering intensity can be represented as a sum of Hankel transforms of the corresponding orders from the multipole components of ρ_r. This ab-initio approach can be successfully applied in description of low-resolution structures [30].

A combined method which make use of the Debye formula and spherical harmonics leads to scattering intensity of the form:

$$I(q) = 2\pi^2 \sum_{r=0}^{\infty} \sum_{m=-l}^{l} \left\{ \sum_{j=1}^{k} \left[\Delta\rho_j A_{lm}^{(j)}(q) \right]^2 + 2 \sum_{n>j} \Delta\rho_j A_{lm}^{(j)}(q) \Delta\rho_n \left[A_{lm}^{(n)}(q) \right]^* \right\}, \tag{3.91}$$

where K is the number of phases of a particle and the symbol $*$ denotes the complex conjugate. This expression significantly speeds up the calculations by considering that the search space is filled with spherical beads. Eq. (3.91) is implemented in the program DAMMIN [31], which makes use of a heuristic optimization to select an appropriate subset of this set of spherical beads in order to generate the best fit for the experimental data.

3.9.4 Fast Fourier Transform

This is a more recent method suggested in Ref. [32] to calculate the scattering intensity by using the Fast Fourier Transform of a 3D model, defined on a cubic lattice of dimension Na, with N^3 points and with SLD $\rho(r)$ spaced by the constant lattice a. The scattering amplitude is the discrete Fourier of this distribution, and the intensity of a basic unit becomes:

$$I(q) = \mathscr{F}\{\rho(r)\}^2 \left(\frac{\prod_{i=1}^{m} \sin q_i a/2}{q_i a/2} \right)^2, \tag{3.92}$$

where the symbol $\mathscr{F}\{\cdots\}$ denotes a discrete Fourier transform, and m is the number of dimensions. The advantage of this approach is the low computational cost, that is the time increase with $\mathscr{O}(n \ln n)$.

3.10   Case Studies and Applications

For illustration purposes we present in this section two case studies and an application, which make use of the previous concepts and methods and which will be also useful in constructing and extracting information from fractal structures.

We consider first determination of structural parameters from two macromolecular systems: lysozyme and glucose isomerase [9]. Figure 3.4a shows the corresponding SAXS intensity (data including the measurement errors), together with the scattering from a sphere of known radius $R = 2$ nm (continuous line), for comparison. The experimental data show the presence of the Guinier region for both systems. This is followed by an intermediate region where the intensity decays proportionally to some power α of the scattering vector, that is $I(a) \propto q^{-\alpha}$. Finally, at high values of the wavevector q, the background is attained. This is seen as a liner-like region, with high vales of errors in the scattering intensity. In the first step we perform a Guinier plot, that is a plot of $\ln I(q)$ versus q^2, where the data show a linear behaviour. Figure 3.4b shows the corresponding curves (discrete data), together with the corresponding linear fit given by Eq. (3.60). The fitting procedure has been performed in the range $0.1 \lesssim q \lesssim 1$ nm^{-1} for lysozyme, and at $0.1 \lesssim q \lesssim 4.6 \times 10^{-1}$ nm^{-1} for glucose isomerase. The results show that in the case of lysozyme $R_g = 1.41$ nm, for glucose isomerase $R_g = 3.38$ nm, while for the sphere $R_g = 1.55$ nm.

Figure 3.5a and b shows the corresponding Kratky plot, that is a plot of $q^2 I(q)$ versus q, and respectively the pair distance distribution function $p(r)$ of the same molecular structures. One can observe that the curves in the Kratky plot have a Gaussian-like shape, which indicates a globular shape without sharp edges for all structures. In the case of glucose isomerase an additional plateau is present at high values of the wavevector q, and which indicates a partially unfolded molecular structure.

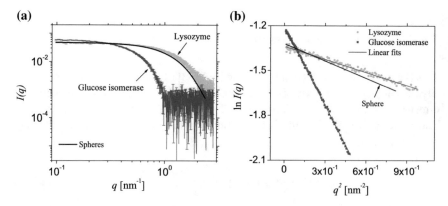

Fig. 3.4 a SAXS intensity and Guinier plot from lysozyme and glucose isomerase (data including the measurement errors), and from a sphere (continuous line) with radius $R = 2$ nm. **b** The corresponding Guinier plot $\ln I(q)$ versus q^2. Raw data from BioXSTAS RAW package described in Ref. [33]

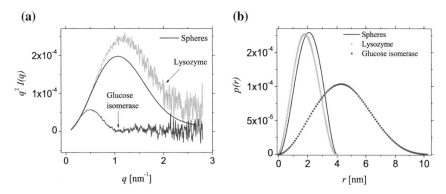

Fig. 3.5 a Kratky plots and **d** Pair distance distribution functions of lysozyme, glucose isomerase (data including the measurement errors) and a sphere (continuous line) with radius $R = 2$ nm. Raw data from Bio*XSTAS RAW* package described in Ref. [33]

The corresponding distribution functions have also a Gaussian-like structure with peaks situated at $r \approx 1.85$ nm for lysozyme, $r \approx 2.02$ nm for sphere and $r \approx 4.2$ nm for glucose isomerase. These values indicate the maximum number of distances inside each structure. The distribution function $p(r)$ takes zero values at $r \approx 4$ nm for both lysozyme and sphere, and at $r \approx 10.2$ nm or glucose isomerase, corresponding to the maximum diameters inside each structure.

By using SAXS data in Fig. 3.4a or from the intercept of the fitting function given by Eq. (3.60) one can determine also the intensity at zero angle, i.e. $I(0)$. This quantity is especially useful when we are interested in determination of the molecular weights (see Eq. (3.61)). In the case of the lysozyme studied here, we have $c_m = 4.27$ mg/ml and $I(0)_m = 0.046$ cm^{-1}. Thus, the obtained molecular weight is $MW = 14.4$ kDa (see Ref. [33] for details concerning the standards). In a similar manner, for glucose isomerase with $c_m = 0.4$ mg/ml and $I(0)_m = 0.061$ cm^{-1} one find that $MW = 173.7$ kDa [33].

Finally, the overall shapes of the molecules are determined by using the method of spherical harmonics presented before, and which is implemented in the ATSAS suite [34]. Thus, starting from experimental data shown in Fig. 3.4a, the most probable configuration of the lysozyme and glucose isomerase molecules are shown in Fig. 3.6a, and respectively in Fig. 3.6b. The results show that the lysozyme has a slightly flattened globular structure, while the glucose isomerase has a more spherical symmetry and with a slightly more rough surface. These confirm the initial results suggested before from the Kratky analysis, and according to which, both molecules have a globular-like structure.

As an application of the Monte Carlo method presented in the previous section, we illustrate determination of the pair distance distribution functions. This approach is very useful when we have to model SAS data for very complex structures and for which $p(r)$ and $I(q)$ are not known analytically. In the following we consider

Fig. 3.6 3D reconstruction using dummy atom models with DAMMIF program of: **a** Lysozyme with maximum diameter 4 nm, and **b** Glucose isomerase with maximum diameter 10.2 nm. The images show the most probable configuration, since the reconstruction is not unique. Raw data from Bio*XSTAS RAW* package described in Ref. [33]

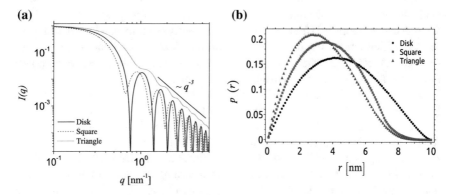

Fig. 3.7 Scattering intensities **a** and pair distance distribution functions **b** of a disk with radius 5 nm (continuous black line), a square (dashed red line) and an equilateral triangle (dotted blue line) with edges 10 nm. For 2D structures the Porod exponent is –3

three basic shapes: a disk of radius $R = 5$ nm, a square of edge $L = 10$ nm and an equilateral triangle of edge $a = 10$ nm.

By using the scattering intensities listed in Table 3.2, the scattering intensities of the disk, square and a triangle, are shown in Fig. 3.7a. As expected, all the features discussed for cubes and spheres can be seen also in Fig. 3.7a, with the main difference that the exponent of the Porod region is –3, which is a signature of scattering from 2D structures. Thus, by using Eq. (3.51), the $p(r)$ functions shown in Fig. 3.7b are obtained. Note that the distribution function of a triangle has the smallest symmetry, and it reaches the highest values faster as compared with pddf of a square and of a disk.

References

1. Feigin, L.A., Svergun, D.I.: Structure Analysis by Small-Angle X-Ray and Neutron Scattering. Springer, Boston (1987)
2. Blüegel, S.: Scattering Methods for Condensed Matter Research: Towards Novel Applications at Future Sources. Forschungszentrum Jülich GmbH, Julich (2012). https://juser.fz-juelich.de/record/136382?ln=de
3. Melnichenko, Y.B.: Small-Angle Scattering from Confined and Interfacial Fluids. Springer International Publishing, Berlin (2016)
4. Goodman, J.W.: Introduction to Fourier Optics. Roberts & Co (2005)
5. Hecht, E.: Optics, 5th edn. Pearson (2016)
6. Melnichenko, Y.B., Wignall, G.D.: J. Appl. Phys. **102**(2), 021101 (2007)
7. Glatter, O., Kratky, O.: Small Angle X-ray Scattering. Academic, New York (1982)
8. Glatter, O., May, R.: International Tables for Crystallography, pp. 89–112. International Union of Crystallography (2006)
9. BioXTAS RAW BioXTAS RAW 1.5.0 documentation. https://bioxtas-raw.readthedocs.io/en/latest/index.html
10. Mylonas, E., Svergun, D.I.: J. Appl. Cryst. **40**(s1), s245 (2007)
11. Rambo, R.P., Tainer, J.A.: Nature **496**(7446), 477 (2013)
12. Fischer, H., de Oliveira Neto, M., Napolitano, H.B., Polikarpov, I., Craievich, A.F.: J. Appl. Cryst. **43**(1), 101 (2010)
13. Stuhrmann, H.B.: J. Appl. Cryst. **7**(2), 173 (1974)
14. Lindner, P.P., Zemb, T.T.: Neutrons, X-rays, and Light: Scattering Methods Applied to Soft Condensed Matter. Elsevier, Amsterdam (2002)
15. Cherny, A.Y., Anitas, E.M., Osipov, V.A., Kuklin, A.I.: Phys. Rev. E **84**(3), 036203 (2011)
16. Anitas, E.M., Slyamov, A., Todoran, R., Szakacs, Z.: Nanoscale Res. Lett. **12**(1), 389 (2017)
17. Schmidt, P.W.: J. Appl. Cryst. **24**(5), 414 (1991)
18. Glatter, O.: IUCr. J. Appl. Cryst. **10**(5), 415 (1977)
19. Weyerich, B., Brunner-Popela, J., Glatter, O.: IUCr. J. Appl. Cryst. **32**(2), 197 (1999)
20. Goodisman, J.: J. Appl. Cryst. **13**(2), 132 (1980)
21. Cherny, A.Y., Anitas, E.M., Osipov, V.A., Kuklin, A.I.: Phys. Chem. Chem. Phys. **19**(3), 2261 (2017)
22. Debye, P.: Ann. Phys. **46**, 809 (1915)
23. Pantos, E., van Garderen, H.F., Hilbers, P.A., Beelen, T.P., van Santen, R.A.: J. Mol. Struct. **383**(1–3), 303 (1996)
24. Putnam, D.K., Weiner, B.E., Woetzel, N., Lowe, E.W., Meiler, J.: Proteins **83**(8), 1500 (2015)
25. Anitas, E.M., Slyamov, A.: PLOS ONE **12**(7), e0181385 (2017)
26. Burley, S.K., Berman, H.M., Bhikadiya, C., Bi, C., Chen, L., Costanzo, L.D., Christie, C., Duarte, J.M., Dutta, S., Feng, Z. et. al.: Nucleic Acids Res. **47**(D1), D520 (2019)
27. Svergun, D., Barberato, C., Koch, M.H.J.: J. Appl. Cryst. **28**, 768 (1995)
28. Kaya, H.: IUCr. J. Appl. Cryst. **37**(2), 223 (2004)
29. Stuhrmann, H.B.: Acta Cryst. A **26**(3), 297 (1970)
30. Svergun, D.I., Volkov, V.V., Kozin, M.B., Stuhrmann, H.B.: Acta Cryst. A (1996)
31. Svergun, D.: Biophys. J. **76**(6), 2879 (1999)
32. Schmidt-Rohr, K.: J. Appl. Cryst. **40**(1), 16 (2007)
33. Hopkins, J.B., Gillilan, R.E., Skou, S.: J. Appl. Cryst. **50**(5), 1545 (2017)
34. Franke, D., Petoukhov, M.V., Konarev, P.V., Panjkovich, A., Tuukkanen, A., Mertens, H.D.T., Kikhney, A.G., Hajizadeh, N.R., Franklin, J.M., Jeffries, C.M., Svergun, D.I.: J. Appl. Cryst. **50**(4), 1212 (2017)

Chapter 4
Small-Angle Scattering from Fractals

Abstract The theory of SAS considered in the previous chapter can be extended to investigations concerning structural properties of more complex systems, such as fractals and multifractals. In this chapter we focus on interpretation of SAS intensities and on extracting the structural information about fractal structures based on the behaviour of their scattering curves and the pair distance distribution functions. Basically, almost every SAS analysis of fractal structures presented in the literature is performed either in terms of mass fractals or as surface fractals. After presenting the general theory of scattering from mass and surface fractals, we present and discuss more general models, suitable for analysis of hierarchical systems or multifractals.

4.1 Introduction

As we saw in the previous chapter, various structures give rise to various types of behaviour of SAS intensities. Here, we shall present and discuss the most common situations which are encountered or we may expect in experimental SAS data. This may help to infer a basic geometrical model from the behaviour of the intensity curve, and to identify the fractal class to which the object belong.

We also have learned that fractals are defined as objects having the property of scale-invariance, that is they remain unchanged under a transformation of scale. Although this definition provides a general description, it does not encompass the salient features which would allow us to distinguish between various classes of fractals with similar characteristics. This is in part due to the fact that scale-invariance can generate both self-similar as well as self-affine structures. In both cases, the generated structures can be written as a union of rescaled copies of themselves, but in the former case rescaling is isotropic (i.e. is uniform in all directions), while in the later one the rescaling is anisotropic (i.e. it depends on the direction). Thus, well known fractals such as the Cantor set, Menger sponge, Sierpinski gasket or Koch snowflake are self-similar, while structures generated using fractional Brownian motion or fractional Gaussian noise are self-affine.

Furthermore, self-similar structures can be further divided into exact and statistical self-similar fractals. Thus, the regular Cantor set is exact self-similar, while changing

© The Author(s), under exclusive license to Springer Nature Switzerland AG 2019   65
E. M. Anitas, *Small-Angle Scattering (Neutrons, X-Rays, Light)*
from Complex Systems, SpringerBriefs in Physics,
https://doi.org/10.1007/978-3-030-26612-7_4

randomly the positions of the basic units at each iteration, leads to a statistical fractal of the same fractal dimension. In addition, the fractal dimension can vary with the iteration number, and depending on the exact type of variation, it may lead to fat fractals. Finally, one may distinguish between mass and surface fractals. As we shall see in the following, each different class has its own "fingerprint" on the SAS curve. Thus, the main aim of this chapter is to understand how we can differentiate between various types of fractal based on information provided by the SAS scattering curve.

4.2 Mass Fractals

4.2.1 Statistically Self-similar

We have seen in Fig. 3.2 that the scattering exponent of a 3D object is −4, for 2D structures is −2, while for 1D structures is −1. However, in many SAS experimental data, the scattering exponents take non-integer values, and the question which arise is how to interpret such values? A first attempt in this sense has been performed by Schmidt in Ref. [1], which suggested that the non-integer values arise from a power-law polydispersity of scatterers. Soon after, the connection with fractal systems has been performed, and the scattering exponents have been explained in the framework of fractal geometry [2, 3].

As it was already discussed in Chap. 1, the dimension D_m of a mass fractal, characterizes the manner in which the mass M scales with the radius r of an imaginary sphere centred on the fractal, i.e. $M \propto r^{D_m}$. Similarly, for a surface fractal, the dimension D_s characterizes the manner in which the surface area S scales with changing the resolution r^2, i.e. $S \propto r^{2-D_s}$. To describe the spatial correlations inside of a fractal with size l composed of basic units of size a, one measures the corresponding "mass" (mass, number of particles, volume etc.) for various values of the radius r. Then, the spatial correlations are described by the pair distribution/correlation function $g(r)$, which can be expressed in terms of the pair distance distribution function $p(r)$ defined by Eq. (3.51), through [4]:

$$g(r) = \frac{p(r)V_{fr}}{4\pi r^2},$$ (4.1)

where V_{fr} is the volume of the fractal with size l. Then, the deviations of the particle density n from its mean value at distance r can be written as [3]:

$$n\left[g(r) - 1\right] = \frac{D_m}{4a^{D_m}}r^{D_m-3}e^{-r/l},$$ (4.2)

where a is the radius of the basic units composing the fractal. The corresponding structure factor is obtained by performing a Fourier transform of Eq. (4.2), which leads to:

$$S(q) \simeq 1 + \frac{D_m \Gamma (D_m - 1)}{(qa)^{D_m} \left[1 + (ql)^{-2}\right]^{(D_{m-1})/2}} \sin\left[(D_m - 1)\arctan(ql)\right], \qquad (4.3)$$

where $\Gamma(\cdot)$ is the gamma function. This equation can be used to obtain the fractal radius of gyration as well as the value of the scattering exponent in SAS intensity. Thus, in the limit $q \to 0$, Eq. (4.3) reduces to:

$$S(q) \simeq 1 + \Gamma(D_m + 1)(l/a)^{D_m}\left[1 - (ql)^2 D_m (D_m - 1)/6\right], \qquad (4.4)$$

from which we can see that the radius of gyration is $R_g = D_m (D_m + 1) l^2/2$. In the limit $q \to \infty$, Eq. (4.3) becomes:

$$S(q) \simeq 1 + \frac{D_m \Gamma (D_m - 1)\sin\left[(D_m - 1)\pi/2\right]}{(qa)^{D_m}}. \qquad (4.5)$$

Equation (1.2) is recovered from Eq. (4.5) by taking into account the fact that the fractal signature is observed only when $l^{-1} \ll q \ll a^{-1}$. In practice, D_m is obtained from a double logarithmic plot of SAS curve as a function of momentum transfer, while the size of the fractal building blocks together with its overall size can be estimated from the deviations of the SAS intensity from a straight line. In particular, if the measured scattering exponent τ is smaller than the Euclidean dimension of the space in which the fractal is embedded, then $\tau = D_m$ (see discussion in Chap. 1). Figure 4.1a shows a transmission electron microscopy (TEM) of gold nanoparticles with diameters of about 6 nm, which form clusters with sizes up to about 150 nm. Figure 4.1b shows the corresponding SAS curve of a macroscopic volume, with $0.1 \lesssim q \lesssim 3.5$ nm^{-1}. In this range the scattering intensity shows a power-law behaviour with the absolute value of the scattering exponent $\tau \simeq 2.35$. Since the system is embedded in a 3D Euclidean space, and $\tau < 3$, it turns out that gold nanoparticles form mass fractal structures with fractal dimension $D_m \equiv \tau \simeq 2.35$. The end of the fractal region is at $q_{max} \simeq 1$ nm^{-1} which indicates that the average size of the particles is $2\pi/q_{max} \simeq 6$ nm, which is in agreement with TEM data. At $q \simeq 1.2$ nm^{-1} the curve begins to be dominated by the incoherent scattering. However, at small values of q we can see that the scattering curve does not yet reaches the Guinier region, which indicates that the sizes of the fractal gold nanoclusters is at least $D_{max} \simeq q_{min} = 2\pi/0.1 \simeq 62.8$ nm, and which reflects the overall size given by the TEM image in Fig. 4.1a. Generally, depending on the experimentall conditions and on the sizes of the inhomogeneities, a SAS experiment may show the presence of both Guinier and Porod regions, especially when combined with ultra SAS (USAS), and thus more complete information can be extracted about the investigated system.

4.2.2 Exact Self-similar

Generally, a deterministic mass fractal is constructed from objects (also known as building blocks, basic/scattering units, or simply—units) of the same size and shape.

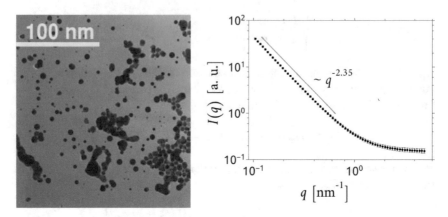

Fig. 4.1 Left side: TEM from gold nanoclusters. Right side: The corresponding SANS curve

If the number of such units is N, each one with form factor $F_0(qa)$, then the fractal form factor can be written as [4]:

$$F(q) = \rho_q F_0(ql)/N, \tag{4.6}$$

where l is the size of the unit, $\rho_q = \sum_{i=1}^{N} \exp(-i\boldsymbol{q} \cdot \boldsymbol{r}_i)$ is the Fourier component of the density of units centers, and $\boldsymbol{r}_i$ are the center-of-mass positions of the units. Thus, by inserting Eq. (4.6) into Eq. (3.68), the scattering intensity becomes:

$$I(q) = I(0)S(q)|F_0(ql)|^2/N, \tag{4.7}$$

where we have (see Eq. (3.68)):

$$I(0) = c|\Delta\rho|^2 V^2. \tag{4.8}$$

Here, V is the fractal volume, and the quantity

$$S(q) \equiv \langle \rho_q \rho_{-q} \rangle / N, \tag{4.9}$$

is the structure factor. It carries information about the relative positions of the units inside the fractal, and is usually expressed as:

$$S(q) = \frac{1}{N} \sum_{i,j=1}^{N} \langle e^{-i\boldsymbol{q} \cdot (\boldsymbol{r}_i - \boldsymbol{r}_j)} \rangle. \tag{4.10}$$

According to Eq. (4.3), it follows that $S(0) = N$, while in the limit of large values of q we have $S(q) \simeq 1$. The last relation holds, since the contribution of non-diagonal terms in Eq. (4.10), i.e. $\sum_{i \neq k} \langle \exp(i\boldsymbol{q} \cdot (\boldsymbol{r}_i - \boldsymbol{r}_j)) \rangle / N$, tends to zero at high values of

q due to the randomness of the phase. Note that the choice of the form and structure factors in Eq. (4.7) is rather arbitrarily and involves a regrouping of terms in the total scattering amplitude $A(\boldsymbol{q})$. In particular, if the quantity $F_0(q)$ is chosen to describe the fractal as a whole, then the structure factor will represent the spatial correlations between different fractals. However, since the fractal positions are assumed to be completely uncorrelated, $S(q) \simeq 1$ at $q \neq 0$.

For deterministic fractals, direct application of Debye formula, Monte Carlo simulations or multipole expansion (for globular-like fractals) in calculating the scattering intensity given by Eq. (4.7) can be a very time-consuming task for iterations exceeding $m = 4$ or 5, since the number of units forming the fractal increases exponentially with the iteration number. However, for symmetric deterministic fractals we can easily overcome this issue by performing a Fourier transform of the underlying fractal measure, and obtain an analytic expression for calculating the fractal form and structure factors.

4.2.3 Application to SAS from Cantor Mass Fractals

We illustrate this approach for the well known 2D Cantor fractal, and afterwards we will extend our approach to more complex systems, such as Sierpinski triangles (gaskets). The 2D Cantor fractal can be built by starting from a homogeneous square of edge size a, which is called the initiator ($m = 0$) (see Fig. 4.2). We choose a Cartesian coordinates system with the origin in the center of the square, and with axes parallel to the square edges. At first iteration ($m = 1$) we divide the initiator into 9 smaller squares of edge length $a/3$. This is called the generator, and in order to obtain the 2D Cantor fractal, we keep only the squares in the corners, as shown in Fig. 4.2. In order to obtain other types of fractals we have to choose a different generator. For example the 2D Vicsek fractal is obtained by keeping the additional square in the center of the initiator, together with the squares from the corners [4]. At the second iteration ($m = 2$) the procedure is repeated for all the 4 squares of edge $a/3$, thus giving 16 squares of edge length $a/3^2$. Higher iterations are obtained in the same manner. The Cantor fractal is obtained in the limit $m \to \infty$ and is composed of $k_m = 4^m$ squares with edge length $a_m = a/3^m$, where m is called the fractal iteration number. Therefore, by using the apparent relation given by Eq. (2.20) $k_m \simeq (a/a_m)^D$, we can see that the fractal dimension is given by:

$$D \equiv \lim_{m \to \infty} \frac{\ln k_m}{\ln (a/a_m)} \simeq 1.26. \tag{4.11}$$

As we can see from its construction (Fig. 4.2), the 2D Cantor fractal is the direct product of two 1D Cantor fractals. Thus, an arbitrarily point $\boldsymbol{r} = (x, y)$ belongs to the 2D Cantor fractal if and only if both x and y belong to the 1D Cantor fractal. The fractal dimension given by Eq. (4.11) is two times larger than that of the 1D

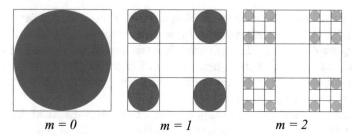

Fig. 4.2 First three iterations $m = 0$ (initiator), $m = 1$ (generator) and $m = 2$ of the 2D Cantor fractal. The radii of the circles are: $a_0 = a/2$, $a_1 = \beta a_0$ and $a_2 = \beta^2 a_0$

Cantor fractal. The shape of the initiator can be chosen arbitrarily, without affecting the value of the fractal dimension, as much as the scaling factor is kept at $\beta = 1/3$.

As a first step in obtaining the fractal form factor, we determine the relative positions of the squares (units) inside the fractal, and express them with the help of a function in reciprocal space, which is known as the generative function of the fractal and is denoted by $G_m(\boldsymbol{q})$. The index m indicates that the generative function depends on the iteration number. For the Cantor fractal shown in Fig. 4.2, the corresponding generative function at $m = 1$ can be written as:

$$G_1(\boldsymbol{q}) = \cos\left(q_x a\beta\right) \cos\left(q_y a\beta\right). \tag{4.12}$$

In order to obtain the fractal form factor at $m = 1$, we make use of the following properties of the form factor of an arbitrarily-shaped particle:

1. $F(\boldsymbol{q}) \rightarrow F(\beta\boldsymbol{q})$ when the particle is scaled as $a \rightarrow \beta a$,
2. $F(\boldsymbol{q} \rightarrow) F(\boldsymbol{q}) e^{-i\boldsymbol{q}\cdot\boldsymbol{a}}$ when the particle is translated as $\boldsymbol{r} \rightarrow \boldsymbol{r} + \boldsymbol{b}$,
3. $F(\boldsymbol{q}) = \left[A_{\mathrm{I}} F_{\mathrm{I}}(\boldsymbol{q}) + A_{\mathrm{II}} F_{\mathrm{II}}(\boldsymbol{q})\right] / (A_{\mathrm{I}} + A_{\mathrm{II}})$ when the particle consists of two non-overlapping subsets I and II.

At $m = 1$, the center positions of the four squares are shifted from the center by vectors $\boldsymbol{b}_j = \{\pm\beta a, \pm\beta a\}$ with various combinations of the signs. Then, by using the above properties, together with the form factor of a disk of unit radius, given by:

$$F_0(q) = 2J_1(q)/q, \tag{4.13}$$

where $J_1(q)$ is the Bessel function of the first kind, we can write:

$$A_1 F_1(\boldsymbol{q}) = \sum_{j=1}^{4} \beta^2 A_0 F_0(\beta q a/2) e^{-i\boldsymbol{q}\boldsymbol{b}_j}, \tag{4.14}$$

where $A_1 = k_1 A_0 \beta^2$ is the area of Cantor fractal at $m = 1$, and $A_0 = \pi(a/2)^2$ is the area at $m = 0$. Rewriting explicitly the last equation, one obtains:

$$F_1(q) = G_1(q)F_0(\beta qa/2). \tag{4.15}$$

The form factor at $m = 2$ can be obtained by the same procedure, which yields:

$$F_2(q) \equiv G_1(q)F_1(\beta q) = G_1(q)G_1(\beta q)F_0(\beta^2 qa/2). \tag{4.16}$$

For an arbitrarily fractal iteration number, we can write the following relation:

$$F_m(q) = F_0(\beta^m qa/2) \prod_{i=1}^{m} G_i(q), \tag{4.17}$$

where, by definition we choose $G_0(q) = 1$. Therefore, at arbitrarily iteration number m the scattering intensity can be written as:

$$I_m(q)/I_m(0) = \langle |F_m(q)|^2 \rangle, \tag{4.18}$$

which is a function of qa and the scaling factor β.

Equations (4.6) and (4.17) show that the Fourier component of the density of square centers at mth iteration are given by:

$$\rho_q^{(m)} = k_m \prod_{i=0}^{m} G_i(q), \tag{4.19}$$

which can be used in Eq. (4.9) to obtain:

$$S_m(q)/k_m = \left\langle \prod_{i=1}^{m} |G_i(q)|^2 \right\rangle. \tag{4.20}$$

Finally, by using Eqs. (4.17), (4.18) and (4.20), one obtains:

$$I_m(q)/I_m(0) = |F_0(\beta^m qa/2)|^2 S_m(q)/k_m, \tag{4.21}$$

and which is in agreement with the general relation in Eq. (4.7).

Since the main aim of any SAS experiment is to extract information about the structure in real space from scattering data, we make use here of the pair distance distribution function $p(r)$ introduced previously to describe the of the units composing the fractal. For this purpose, we consider first that the fractal is monodisperse. At a given iteration number, the 2D Cantor fractal consists of squares of the same size. Thus, Eq. (4.10) can be rewritten as:

$$S(q) = 1 + \frac{2}{k_m} \sum_{1 \le i \le j \le k_m} \frac{\sin qr_{ij}}{qr_{ij}}, \tag{4.22}$$

where $r_{ij} \equiv |r_i - r_j|$ are the relative distances between square centers. Note that, in deriving the last expression, the property given in Eq. (3.48) has been used. Due to the symmetry of the fractal, many distances separating different points coincide, and thus the sum contains many equal terms. In order to express the probability density of finding the distance r between the centers of two arbitrarily squares inside the fractal, one consider the pddf $p(r)$, such as [4]:

$$p(r) \equiv \frac{2}{k_m (k_m - 1)} \sum_{i<j} \delta \left(r - r_{ij} \right) = \frac{2}{k_m (k_m - 1)} \sum_{r_p} C_p \delta \left(r - r_p \right), \quad (4.23)$$

where C_p are the number of distances separated by r_p. Thus, the structure factor becomes:

$$S(q) = 1 + (k_m - 1) \int_0^\infty p(r) \frac{\sin qr}{qr} dr, \quad (4.24)$$

and its Fourier transform gives:

$$p(r) = \frac{2}{\pi} \int_0^\infty \frac{S(q) - 1}{k_m - 1} qr \sin (qr) \, dq. \quad (4.25)$$

Figure 4.3 shows the monodisperse form (left part) and structure (right part) factor given by Eqs. (4.18) and (4.20) for the first three iterations. Their main feature is the presence of the Guinier region at $q \lesssim \pi/a \simeq \pi$ and a fractal region at:

$$\pi/a \lesssim q \lesssim \pi/a_m. \quad (4.26)$$

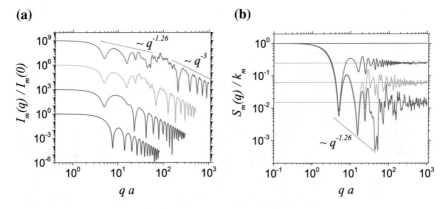

Fig. 4.3 **a** Form factor (Eq. (4.18)) of 2D Cantor fractal for iterations $m = 0, 1, 2, 3$ (from bottom to top). The scattering curves are scaled up by a factor of 10^{3m} for clarity. **b** Structure factor (Eq. (4.20)) of 2D Cantor fractal for iterations $m = 0, 1, 2, 3$ (from top to bottom). The horizontal lines indicate the asymptotic values. Here, k_m is the number of disks at mth iteration, and a is the edge size of the square circumscribing the disk at $m = 0$

However, for $\pi/a_m \lesssim q$ the form factor is characterized by a Porod-law decay with the scattering exponent -3, while the structure factor attains its asymptotic values. In the following we shall discuss each region separately.

In the Guinier region the intensity is well described by the expansion of scattering intensity given by last equality in Eq. (3.56). The intensity at $q = 0$ is given by Eq. (4.8) but with the volume V replaced by the corresponding area, and $S_m(0) = k_m$. By performing a series expansion of the fractal form factor given by Eq. (4.17) in power series of qa, and using Eq. (4.18), analytic expressions for the fractal radius of gyration at arbitrarily iteration number can be obtained. For example, the fractal radius of gyration of the 3D version of the Cantor fractal can be written as [5]:

$$R_g = \left(\beta^{2m} R_{g0}^2 + 3\beta^2 \frac{1 - \beta^{2m}}{1 - \beta^2} a^2 \right)^{1/2}, \tag{4.27}$$

where $R_{g0} \equiv (5/3)^{1/2} a/2$ is the radius of gyration of a uniform ball (see Table 3.2). By considering the idealized version of the fractal, that is the structure in the limit of high number of iterations with $\beta^{2m} \ll 1$, we can rewrite the previous equation such as:

$$R_g = \frac{3^{1/2}\beta a}{\left(1 - \beta^2\right)^{1/2}}. \tag{4.28}$$

The fractal region of a mass fractal is determined by the largest and respectively by the smallest distances between center-of-masses of the composing units. For the 2D Cantor fractal studied here, this corresponds to distances between the center of disks (see Fig. 4.2). Since the distances are of the orders of a and respectively of $a\beta^m$, then their inverses set the limits of the fractal region in the reciprocal space, as discussed above. These estimations are obtained from analytical expressions given by Eqs. (4.17) and (4.18), where at a given q we have $G_{m+1} \simeq 1$ if the cosine argument is much smaller than 1. Note that an increase of the iteration number does not bring us a significant correction. This shows that at mth fractal iteration the corresponding scattering amplitude is the same as that of the ideal fractal in the region $qa\beta^m \lesssim \pi$, and which is in agreement with the upper limit in Eq. (4.26).

Since the form factor of the disk plays a role only when $qa\beta^m \gg \pi$, it turns out that in the fractal region the scattering intensity is very similar to that of the structure factor, that is $I_m(q)/I_m(0) \simeq S_m(q)/k_m$. This can be very clearly seen in Fig. 4.4a where a direct comparison between the form and structure factors is performed for all regions, at $m = 4$. The structure factor is usually interpreted by analogy with optics, that is if at $m = 1$, $k_1 G_1(q)$ is the amplitude produced by the four point-like disks, then the quantity given by $k_1^2 \langle |G_1(q)|^2 \rangle \equiv S_1(q)/k_1$ represents the scattering intensity averaged over all directions of momentum q. The analogues of this situation in optics is the diffraction with an entirely uncollimated beam, and which is the source of strong spatial incoherence. The main effect is that only the first minimum and maximum are clearly observable. Their interpretation in the framework of geometrical optics is discussed in Ref. [4]. For scattering at $m = 1$ the

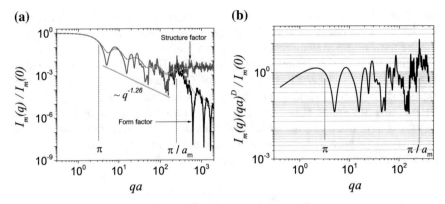

Fig. 4.4 a Monodisperse and polydisperse fractal form and structure factors of the 2D Cantor fractal at $m = 4$ (see Ref. [5]). The vertical dashed lines indicate the limits of the fractal region. The dashed curve (green) is the polydisperse form factor, and the dotted (blue) curve is the polydisperse structure factor, with a log-normal distribution of fractal sizes given by Eq. (3.69). In both cases the relative variance is $\sigma_r = 0.2$. **b** The function $I_m(q)q^D$ is log-periodic in the fractal region, with the period equal to $\log_{10}(1/\beta)$. Here, $\beta = 1/3$ is the fractal scaling factor, $D = 1.26$ is the fractal dimension, while a_m and a are the edge size of the square circumscribing the disks at an arbitrarily iteration $m \neq 0$, and respectively at $m = 0$

main conclusion is that a phase $\phi \lesssim 1$ corresponds to an entirely coherent regime, $\phi \simeq \pi$ and $\phi \simeq 2\pi$ correspond to minimum and maximum, while for $\phi \geq 2\pi$ an entirely incoherent regime is attained. By extending this analysis to an arbitrarily iteration number m, we can conclude that in the fractal region of the structure factor, each of the most pronounced maxima and minima correspond to the interferences of the amplitudes from disks at a given fractal iteration.

Thus, the generalized power-law behavior of the structure factor can be described through the approximation $S_m(q/\beta) \simeq \beta^{D_m} S_m(q)$, which together with Eq. (4.21) and considering that $F_0 \simeq 1$, leads to:

$$I_m(q/\beta)\,(q/\beta)^{D_m} \simeq I_m(q)q^{D_m}, \tag{4.29}$$

that is the scattering intensity is approximately log-periodic, with the period equal to the inverse of the scaling factor $1/\beta$.

The influence of polydispersity on SAS form and structure factors is shown in Fig. 4.4a. As expected the monodisperse scattering curves become smeared out, but nevertheless, the value of the scattering exponent remains unchanged and its absolute value coincides with the fractal dimension, in the fractal region. For the chosen value of the relative variance (i.e. $\sigma_r = 0.2$), and which control the width of the distribution function D_N, the polydisperse curves are still characterized by small oscillations in the fractal region. However, it is known that increasing σ_r can lead to a completely smoothed curve [4, 6, 7]. Depending on the type of fractal this behavior is achieved starting from $\sigma_r \simeq 0.3$. The scattering exponent is preserved also in the Porod region, i.e. it equals -3 for the form factor, and is 0 for the structure factor.

Figure 4.4b illustrates this property, where the function $I_4(q)q^{D_m}$ versus q corresponding to the 2D Cantor fractal has been plotted on a double logarithmic scale. Note that the upper limit of the fractal region depends also on the fractal iteration number m. Thus, in addition to the scaling factor β the iteration number can be used as a parameter to set up the length of the fractal region in momentum space. In addition, instrumental effects and polydispersity can smear the scattering curve such that a simple power-law decay, similar to those given by random fractals is obtained. Thus, in order to distinguish between random and deterministic fractals from SAS experimental data, the measurements have to be performed at instruments with very good resolution function, and with samples that contain fractals with small values of the relative variance describing their size distribution.

In the Porod region (i.e. for $q \gtrsim \pi / (\beta^m a)$) the structure factor tends to 1, and therefore, by using Eq. (4.21) the scattering intensity becomes:

$$I_m(q)/I_m(0) = |F_0(\beta^m qa)|^2/k_m. \qquad (4.30)$$

Therefore, in the Porod region the scattering intensity is given by the intensity of the basic units (i.e. disks) and the corresponding maxima obey the Porod's law. A quite frequently case occurs when the size of the basic units are much smaller than the distances between them [8]. As a consequence, an intermediate region appears, where $I_m(q)/I_m(0) \simeq 1/k_m$. Therefore, in between the fractal and Porod regions a "shelf" appears near $1/k_m$. This case will be discussed in more details later in the context of SAS from deterministic surface fractals.

The properties of the 2D Cantor fractal in the real space can be inferred from Fig. 4.5, where the coefficients in Eq. (4.23) are shown for the iteration numbers

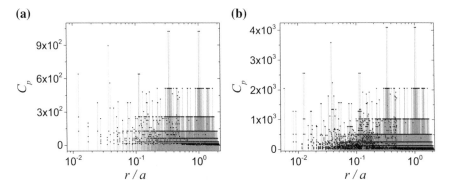

Fig. 4.5 The coefficients C_p in Eq. (4.23) for the pair distance distribution function of the 2D Cantor fractal at fractal iterations $m = 5$ (**a**) and at $m = 6$ (**b**). The positions of the centers of groups are the most common distances between the disks. The log-periodicity is $\log(1/\beta)$, where $\beta = 1/3$ is the scaling factor. Here, a is the edge size of the square circumscribing the disk at iteration $m = 0$

$m = 5$ and $m = 6$. Their values are found via a simple numerical procedure. The main property is the presence of groups of maxima and minima, with the most visible ones being situated at $r/a \simeq 0.3$ and at $r/a \simeq 1$. Note that for 3D fractal structures, where at a given iteration the number of distances is much bigger than for 2D fractals, the groups of distances are more clearly distinguished (see Ref. [4]), and thus the value of the scaling factor can be determined more precisely. The centers of these groups, which represent distances between clusters of disks (see discussion above), show a periodicity on the logarithmic scale, with the period related to the fractal scaling factor through $\log (1/\beta)$. This periodicity is a signature of the self-similarity property of fractals. When the polydispersity of the fractals is taken into account, the pair distance distribution function can be calculated according to:

$$\langle p(r) \rangle_D = \frac{2}{k_m (k_m - 1)} \sum_p \frac{C_p}{z_p} D_N \left(\frac{r}{z_p} \right), \tag{4.31}$$

where $z_p = r_p / a$. This equation is very efficient to be used at high number of iterations, when C_p and z_p are known, since it avoids the drawback of calculating an exponentially increasing number of distances. The effect of introducing the polydispersity is similar to that on scattering curves, in the reciprocal space, that is the $p(r)$ is smeared out.

Since the pair distribution function $g(r)$, which describes the spatial correlations between the disks inside the fractal, satisfies $g(r) \propto r^{D_m - 3}$ for $r \lesssim a$, and taking into account Eq. (4.1), it follows that $p(r) \propto r^{D_m - 1}$. The generalized power-law behaviour can be inferred from the behaviour of C_p shown in Fig. 4.5. Since in the fractal region the structure factor satisfies the property $S_m(q/\beta) \simeq \beta^{D_m} S_m(q)$, the quantity $p(r)/r^{D_m - 1}$, is also log-periodic in the fractal region. From another hand we can see that the fractal region in the reciprocal space represents the main contribution in the integral of Eq. (4.25). Since $S(q) \gg 1$, one can make the substitutions $S(q) - 1 \to S(q)$ and $q \to q/\beta$, which leads to [4]:

$$\frac{p(\beta r)}{(\beta r)^{D_m - 1}} = \frac{p(r)}{r^{D_m - 1}}, \tag{4.32}$$

which holds true for $\beta^{m-1} a \lesssim r \lesssim a$. In a similar manner, we can derive equivalent relations for the pair distribution function $g(r)$.

The coefficients C_p given by Eq. (4.23) and shown in Fig. 4.5 describe the distances between the centers of the disks composing the 2D Cantor fractal, that is the disks are assumed to be point-like particles. For cases when such an assumption can not be used, we may either calculate analytically the corresponding correlation function $\gamma(r)$ of the fractal followed by determination of $p(r)$ (pddf), or we can simulate the structure of the fractal using a random arrangement of points found inside the region delimited by the disks composing the fractal, followed by a numerical procedure to calculate the corresponding pddf. Since analytic expressions can be quite

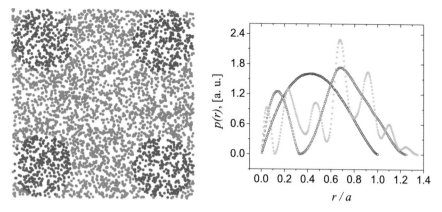

Fig. 4.6 Left part: Approximating the first iteration ($m = 1$) of 2D Cantor set with randomly generated points (red). Right part: the pair distance distribution functions for $m = 0$ (blue), $m = 1$ (red) and $m = 2$ (green) of the 2D Cantor fractal simulated using randomly generated points. Here, a is the edge size of the square circumscribing the disk at iteration $m = 0$

hard to derive for such cases, one make use of the second approach which relies on simple combinatorial analysis.

Figure 4.6 left part shows the simulated 2D Cantor fractal at $m = 1$ (red points). Initially the points are generated randomly on the whole square bounding the fractal. Then, in calculating the pddf, the points found inside the disks corresponding at $m = 1$ are kept, while the others (grey) are discarded. For the simulations performed at $m = 1$, a number of about 10^4 points have been generated. A similar procedure can be applied for arbitrarily iteration number. The calculated pddf for the first three iterations (i.e. $m = 0, 1, 2$) are shown in Fig. 4.6 right part. The number of generated point increases with m and is chosen in such a way that the total number of distances at each m is constant. This increase is necessarily since the corresponding area covered by the disks relative to the area of the square bounding the disk at $m = 0$ is smaller. Here, their number were chosen in such a way that the total number of distances is about 2×10^7. Finally, the pddf has been normalized such that a Fourier transform would give $I(0) = 1$. At $m = 0$ the pddf corresponds to the one of a single disk of radius $1/2$, as expected, with a single minima at $r/a \simeq 0.43$. However, at $m = 1$ the pddf has a minima at $r/a \simeq 0.33$ and two maxima at $r/a \simeq 0.13$ and $r/a \simeq 0.68$. We know that at $m = 1$ the size of the disks are 0.33, and thus the range $0 \leq r/a \leq 0.33$ corresponds to distances inside these disks. For $0.33 \lesssim r/a$, the pddf corresponds to the distances between disks, with the most common ones at $r/a \simeq 0.68$. In this range the shape of the pddf is triangular-like with small oscillations on the right "edge" and which reflects the symmetry of the disk's position inside the fractal. At $m = 2$ the pddf shows a succession of maxima and minima. The first minimum occurs at $r/a \simeq 0.11$ and it one third smaller than the position of minimum at $m = 1$. Thus, their positions reflect the values of the fractal scaling factor. At high values of r/a the pddf decays through a succession of more pronounced maxima and minima (as compared with $m = 1$), thus reflecting the correlations between various clusters of disks. At higher iterations number, additional maxima and minima shall be observed, with their positions related to the fractal scaling factor. Note that the highest values where pddf takes zero values, i.e. $r/a \simeq 1$ for $m = 1$, $r/a \simeq 1.25$ for $m = 2$, and $r/a \simeq 1.36$ at $m = 2$, increase with the iteration number. This shows that their overall size increase, and which is in agreement with the suggested model: at $m = 1$ the maximum distance between the four disks is bigger than the maximum distance within the disk at $m = 1$, and the maximum distance between disks at $m = 2$ is bigger than the maximum distance between disks at $m = 1$. An alternative way to express this effect from SAS experimental data is through the corresponding radii of gyration R_g. For example, for 3D Vicsek fractals or fat fractals, analytical expressions have been derived, and which are in agreement with the present observations [4].

4.2.4   Case Study: Sierpinski Triangles on Au(100)-(hex) Surfaces

As we already saw in Chap. 1, in recent years several methods have been developed to build and characterize deterministic fractal structures at nano- and micro-scales. In this section we focus on SAS analysis of Sierpinski triangles (ST) of the fifth order. The approach presented here to extract the structural information is based on the theoretical framework developed in the previous section, and it can be easily extend to arbitrarily mass fractal structures of a single scale.

The investigated ST consist of Fe atoms, 4,4"-dicyano-1,1':3'1"-terphenyl (C3PC) and 1,3-bi(4-pyridyl) (BPyB) benzene molecules reconstructed on Au(100)-(hex) surfaces and was fabricated in Ref. [9]. They are prepared in ultra-high vacuum by a combination of templating and co-assembling methods. The first method is used to construct the basic units (building blocks) of the ST. Initially, Fe atoms and C3PC form 1D double chains with metal-organic coordinating ST, and then about $13 \div 26\%$ of C3PC molecules are replaced by BPyB to form the building blocks. Its structure, with dimensions of $r_0 \simeq 7.8$ Å is shown in Fig. 1a upper right of Ref. [9]. Here, we shall model the basic building block with an equilateral triangle of the same edge length and centered on the Fe atom. Then, by using the co-assembly method, ST of various iterations can be obtained. For the fifth order ST investigated here, it's overall size is $a = 500$ Å and the scaling factor $\beta = 1/2$. It is clear that both the dimensions of the overall structure and of the individual balls are appropriate for an experimental SAS analysis of such samples in the fractal region, but not enough to see a significant part of the Porod region since the required experimental range is 0.012 Å$^{-1} \lesssim q \lesssim 0.402$ Å^{-1}, which covers the typical q-range available to many instruments.

The structure obtained in Ref. [9] is modeled here in Fig. 4.7a. Note that due to finite instrument resolution the scattering curves will be partially smeared, in a manner similar to the polydispersity, and a background noise typically arise at high

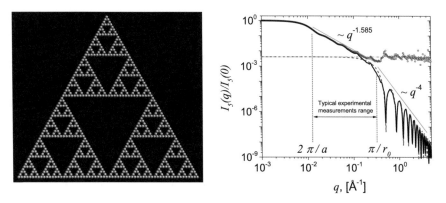

Fig. 4.7 Left part: Model of the ST obtained in Ref. [9], with overall size of $a \simeq 500$ Åand the size of the basic unit (sphere) $2r_0 \simeq 15.6$ Å. Right part: The total scattering intensity (black—continuous line), the form factor of the sphere with radius r_0 (red—dashed line) and the structure factor (blue—circles) of ST

values of momentum transfer q. Thus, the developed model can be used to model the experimental data only after the effects of the resolution function and the background have been minimized and respectively removed.

For the model development, we replace each of the triangles with monodisperse spheres of radii $r_0 \simeq 7.8$ Å. Thus, the system consists of 3D objects (spheres) distributed on a planar surface, with their positions forming a ST, as shown in see Fig. 4.7a. At an arbitrary iteration number m, the ST consists of

$$k_m = 3^m, \tag{4.33}$$

balls of edge size $a_m = a/2^m \sqrt{3}/3$. Thus, the fractal dimension is given by:

$$D = \lim_{m \to \infty} \frac{\log k_m}{\log (a/a_m)} \approx 1.585. \tag{4.34}$$

Thus, the generative functions ($G_i(q)$) of the mth iteration of SG can be written with the help of the product:

$$P_m(q) \equiv \prod_{i=1}^{m} G_i(q), \tag{4.35}$$

where at $m = 1$ the generative function is given by:

$$G_1(q) = \frac{1}{3} \sum_{k=0}^{2} e^{-iq \cdot b_k}, \tag{4.36}$$

and $G_m(\boldsymbol{q}) = G_1(\beta_s^{m-1}\boldsymbol{q})$. Here, $\boldsymbol{b}_k$ are the translation vectors, and are given by:

$$\boldsymbol{b}_k = \frac{a\sqrt{3}}{6} \left\{ \cos\left(\frac{\pi}{3}(2k + \frac{3}{2})\right), \sin\left(\frac{\pi}{3}(2k + \frac{3}{2})\right) \right\}. \qquad (4.37)$$

Thus, by using the vectors $\boldsymbol{b}_k$ from Eq. (4.37) together with the generative function given by Eq. (4.36) and the subsequent property at arbitrarily m and the spherical form factor (see Table 3.2) into Eq. (4.21) one obtains in Fig. 4.7b the corresponding scattering intensity of ST, shown in black—continuous line.

A typical experimental q-range for SANS is $0.007 \text{ Å}^{-1} \lesssim q \lesssim 0.5 \text{ Å}^{-1}$, with possible extensions for experiments performed in ultra-SANS (USANS) mode or by using x-rays. Taking into account the nominal values of the ST dimensions and of its building blocks (spheres), it turns out that the observable signal for this structure will be found in the fractal region, that is where the scattering intensity decays as $I(q) \propto q^{-1.585}$, with the exponent equal to the fractal dimension of ST determined analytically from Eq. (4.34). In addition, from the fractal region we can extract the value of the scaling factor and the fractal iteration number (see details in Fig. 4.8a and subsequent explanations). From the transition points $q \simeq 2\pi/a$ and $q \simeq \pi/r_0$ we can extract information concerning the overall sizes of ST and of the individual spheres, that is $a \simeq 500 \text{ Å}$ and $r_0 \simeq 7.8 \text{ Å}$. At $a \lesssim 2\pi/a$ and $q \gtrsim \pi/r_0$ we have the Guinier and Porod regions (see a more detailed analysis in Fig. 4.8b and c).

Having known the size of the spheres, we can plot the corresponding form factor, which is shown as a red-dashed line in Fig. 4.7b. By using Eq. (4.21), which shows that the scattering intensity can be written as a product of the building block form factor and fractal structure factor, the later one can then be obtained by dividing the scattering intensity to the sphere form factor. As such, the resulting structure factor is shown in Fig. 4.7b in blue—circles. As expected, up to $q \lesssim \pi/r_0$, the structure factor and the scattering intensity coincide (see previous discussion concerning 2D Cantor fractals), while for $q \gtrsim \pi/r_0$ the structure factor reaches its asymptotic values. In Fig. 4.7b they are marked by the horizontal dark yellow—dash-dotted line, and which is situated at about 4.11×10^3. Thus, the number of spheres forming the ST at $m = 5$ is recovered as $1/(4.11 \times 10^3) \simeq 3^5$, in accordance to the model proposed.

Figure 4.8a shows the quantity $I(q)q^{D_m}$, with $D_m = 1.585$ for the fifth iteration of ST. It is clear that the logarithmic periodicity, with period equal to $\log(1/\beta)$, becomes even more pronounced as compared with the curves shown in Fig. 4.7b. It is known that for structure factors, the number of minima is equal to the fractal iteration number. Here, we can see the presence of three periodic minima and a fourth one at $q \simeq 2 \times 10^{-1} \text{ Å}^{-1}$. However, the fourth and fifth minima are influenced by the spherical form factor, and thus their intensities are different. Although the position of the fifth minimum still obeys the log-periodicity, it is not shown here, for clarity.

Figure 4.8b shows a plot of the quantity $\log I(q)$ versus q^2 (Guinier plot) in the Guinier region of ST (black—discrete dots), together with a linear fit (red-continuous line). As discussed in Sect. 3.6, the slope of this curve can be used to estimate the fractal radius of gyration, and which gives $R_g \simeq 6.1 \text{ Å}$. Figure 4.8c shows the

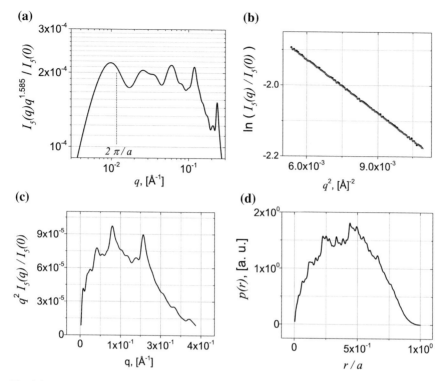

Fig. 4.8 Various representations of SAS data for ST, useful for extracting structural informations. **a** The quantity $I(q)q^{D_m}$ with $D_m = 1.585$ shows an approximate log-periodicity in the fractal region, which is used to obtain the fractal scaling factor. **b** Guinier plot used to estimate the fractal radius of gyration. **c** Kratky plot for assessing the overall configuration. **d** pddf obtained by simulating the structure of ST with Monte-Carlo methods, as for 2D Cantor sets. Here, $a \simeq 500$ Å is the overall size of KS

quantity $q^2 I(q)$ versus q (Kratky plot). Its shape does not show any plateau at high q, excepting a convex decay, thus suggesting a globular-like shape with reduced symmetry. The presence of minima and maxima arise from the internal structure of ST.

Figure 4.8d shows the pddf of ST obtained from Monte Carlo simulations described in Sect. 3.6.2, in units of $a = 500$ Å. The curve has a high degree of symmetry, reflecting the existing ordering inside ST. Note the missing of clear minima reflecting the self-similarity, as in the case of 2D Cantor fractals. We may explain this by considering the relative values of the ratio between sizes of the building blocks l_0 and the distances between them L. While for 2D Cantor sets l_0/L is significantly smaller than one, in the case of ST they are much closer, that is $l_0/L \simeq 1$, and thus in the later case the distances arising from inside the sphere "interfere" with those between them. In turn, This leads to the disappearance of clearly defined minima corresponding to levels of $p(0)$.

4.3 Surface Fractals

4.3.1 Statistically Self-similar

Surface fractals are objects with a rough surface and whose area is given by Eq. (2.3), when measurements are performed with a resolution of r^2. As already pointed out in Sect. 2.2, D_s is the surface fractal dimension, satisfying the condition $d - 1 < D_s < d$, where d is the Euclidean dimension of the space in which the fractal is embedded. If $d = 2$, D_s tends one for almost completely smooth lines, and is equal to two when the curve completely covers the plane (such as a Peano curve). For example, the well known Koch snowflake has $D_s = 1.26$ which shows that the perimeter is quite "wriggled".

In the case of 3D surface fractals, the characteristic correlation function is given by:

$$\gamma(r) = 1 - \frac{1}{4\phi\,(1 - \phi)}\,\frac{v_b(r)}{V}, \tag{4.38}$$

where ϕ is the volume fraction of pores, V is the volume of irradiated sample, and $v_b(r)$ is the volume of a boundary layer. This relation holds in the limit $r \to 0$, and for $r \ll \xi$, where ξ is the correlation length of the surface fractal. By using Eq. (2.23) we can write that:

$$v_b(r) = Sr^{3-D_s}, \tag{4.39}$$

where S is the total area of the sample. Thus, by using Eqs. (3.49), (4.38) and (4.39) one can see that in the limit $\xi q \gg 1$, the corresponding scattering intensity can be written as [10]:

$$I(q) \simeq Kq^{-(6-D_s)}, \tag{4.40}$$

in the range $1/q_{max} < \xi < 1/q_{min}$. Here, q_{max} and q_{min} are the end points for which the surface shows the self-similarity property, while the proportionality constant is given by [10]:

$$K = \pi\,(\rho_m - \rho_s)\,\frac{S}{V}\Gamma(5 - D_s)\sin\left[(3 - D_s)\,\pi/2\right], \tag{4.41}$$

with ρ_m being the SLD of the matrix in which the fractal is embedded, and ρ_s is the SLD of the fractal itself. It is clear that in the case of a two-phase system with smooth interface (i.e. $D_s = 2$) and constant SLD in either phase, the Porod law is recovered.

If one considers that in a porous system, the volume of the void space, rather than that of the mass, obeys the scaling law $V(r) \propto r^{D_p}$, where D_p is the pore fractal dimension, then a more general expression for the scattering intensity of a mass/surface/pore fractal, is given by [11]:

$$I(q) \propto q^{D_s - 2(D_m + D_p) + 6}, \tag{4.42}$$

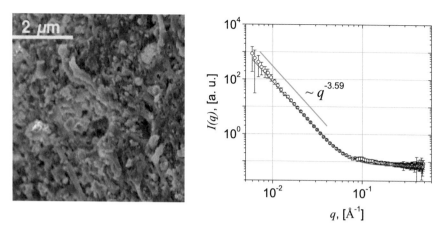

Fig. 4.9 Left part: SEM of a Stomaflex creme polymer matrix with catalyst and Fe particles (see Ref. [12]). Right part: The corresponding SANS spectra (black—discrete points) and a fit with Eq. (4.40) (red—continuous line) plus a constant background term

and which is a particular case of Eq. (1.1), when $d = 3$. When $D_m = D_p = 3$ (surface fractals), Eq. (4.40) is recovered, as a particular case of Eq. (1.3). When $D_s = D_m$ and $D_p = 3$ (mass fractal), Eq. (1.2) is recovered, while for $D_s = D_p$ and $D_m = 3$, Eq. (4.42) reduces to:

$$I(q) \propto q^{-D_p}. \tag{4.43}$$

Note that for two-phase systems, where a transition layer of finite width between the phases has a particular distribution of SLD, the scattering exponent can take values smaller than –4. In particular, for a Gaussian profile of the SLD, the lowest limit of the exponent is –6.

As an application of the above theory, we consider in the following a nanocomposite system consisting of a Stomaflex creme (SC) matrix with catalyst, in which are dispersed Fe microparticles at about 25% volume concentrations. Figure 4.9a shows a SEM image of the SC matrix. The sample was prepared by a mechanical mixing with all the components commercially available. The corresponding SANS curve was registered in the range $6 \times 10^{-3} \text{ Å}^{-1} \lesssim q \lesssim 0.5 \text{ Å}^{-1}$. The main feature of the data is the presence of an almost linear power-law decay up to $q \simeq 5 \times 10^{-2} \text{ Å}^{-1}$ followed by a background component. The experimental data have been modeled throughout the whole range by using Eq. (4.40), to which a constant term, to describe the background has been added. The obtained scattering exponent in the linear region is $\tau \simeq 3.59$, which leads to a fractal dimension of $D_s \simeq 6 - 3.59 = 2.41$. This numerical value corresponds to a surface fractal and reflects the structural modifications which occurs during the blending process and due to the presence of fillers, affecting the interface with the polymer matrix. In terms of the degree of roughness, the composite material is considered to have an average roughness (see discussion in Sect. 2.2). Note that at high q the end of the fractal range is limited by the presence of the incoherent background. In principle, additional procedures can be used to eliminate it, but for the purpose of our example, this is not an issue.

4.3.2 Exact Self-similar

It has been recently shown that any surface fractal can be composed of mass fractals at various iterations, with the same fractal dimensions. Therefore, the scattering amplitude of a surface fractal can be written as a sum of the amplitudes of the composing mass fractals. However, the notion of mass fractal shall be used cautiously here, since in the limit of infinite number of iterations, it can lead to an empty set, and thus it may not exists in the rigorous mathematical sense described in Chap. 1. For practical applications, such as in SAS measurements, this is not an issue since physical structures are always finite, and thus they can not be empty. Therefore, all the scaling properties of the fractal are restricted to a finite range, regardless of whether the limit exists or not at infinite iterations.

Since a surface fractal enables such a decomposition into a sum of mass fractals, in the following we shall derive first, the scattering exponent for the surface fractal intensity given by Eq. (4.40) from the scattering intensity of mass fractals given by Eq. (1.2), at various iterations. For this purpose, we consider a surface fractal an at arbitrarily iteration number m. Figure 4.10 illustrates the construction of a 2D Cantor surface-like (CSF) fractal at $m = 2$. The name comes from the fact that CSF is composed from Cantor mass fractals (CMF) at various iterations. Each individual color (gray level) represents a particular mass fractal iteration. Green (light gray) represent a CMF at $n = 0$, red (gray) corresponds to CMF at $n = 1$, and black (dark) corresponds to CMF at $n = 2$. If we denote with p the number of primary units of size l forming the CMF, and with d the distances between them, then we can write that $p = k^m$, $l \propto \beta^m$, and $d \propto \beta^m$, with $k = 4$ and $\beta = 1/3$ for CMF. In addition, by using Eq. (2.11) with a single scale and with $s = D_m$ and $r = \beta$, one obtains that $\beta = k^{-1/D_m}$.

If we denote by $A(q)$ and $M(q)$ the non-normalized amplitudes of surface, and respectively of mass fractals, then at mth iteration of the surface fractal, one can write [8, 13]:

$$A_m(q) = \sum_{n=0}^{m} M_n(q).$$ (4.44)

From this notation it follows also that the surface fractal intensity is $I_m^{sf}(q) \equiv \langle |A_m(q)|^2 \rangle$ while the mass fractal intensity is $I_m^{(mf)}(q) = \langle |M_n(q)|^2 \rangle$. Then, one can write for the scattering intensities $I_m^{sf}(q)$ that [8, 13]:

$$I_m^{sf}(q) = \sum_{n=0}^{m} I_m^{sf}(q) + \sum_{0 \leq n < p \leq m} \langle M_n^*(q)M_p(q) + M_n(q)M_p^*(q) \rangle,$$ (4.45)

which includes also the correlations between mass-fractal amplitudes. This is a very general relationship, to which several useful approximations can be made.

4.3.3 Neglecting Correlations Between Mass Fractal Amplitudes

By ignoring the second term in Eq. (4.45) and considering only the incoherent sum of mass-fractal amplitudes, we obtain [8, 13]:

$$I_m^{(\text{sf})}(q) \simeq \sum_{n=0}^{m} \langle M_n(q) \rangle . \tag{4.46}$$

Let's consider that at zero-th iteration the volume (or area in 2D case) of the basic unit is V_0, while at arbitrarily mass fractal iteration the volume is $V_n = V_0 \beta^{3n} k^n = V_0 \beta^{n(3-D_m)}$, where $k = 8$ for the 3D version of the Cantor fractal shown in Fig. 4.10. Since $\langle |M_n(q)|^2 \rangle \equiv V_n^2 \langle |F_n^{(\text{mf})}(q)|^2 \rangle$, it turns out from Eq. (4.46) that [13]:

$$I_m^{(\text{sf})}(q) = \sum_{n=0}^{m} V_0^2 \beta^{2n(3-D_m)} \langle |F_m^{(\text{mf})}(q)|^2 \rangle . \tag{4.47}$$

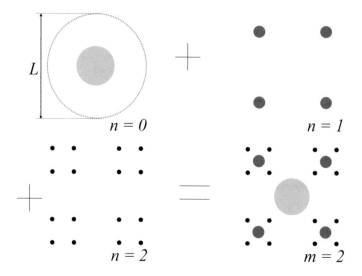

Fig. 4.10 Construction of the CSF at iteration $m = 2$ as a sum of CMF at iterations $n = 0$ (green—light gray), $n = 1$ (red—gray) and $n = 2$ (black—dark) for ratio $d/l \simeq 1$ (see Ref. [8]). L is the overall size of the fractal

It is known that in the case of a mass fractal of fractal dimension D_m, overall size L, and composed of p basic units of size l separated by distances d, with $l \lesssim d \ll L$, the normalized form factor can be written qualitatively as:

$$\left\langle |F^{(mf)}(\boldsymbol{q})|^2 \right\rangle \simeq \begin{cases} 1, & q \lesssim 2\pi/L, \\ (qL/2\pi)^{-D_m}, & 2\pi/L \lesssim q \lesssim 2\pi/d, \\ (d/L)^{D_m}, & 2\pi/d \lesssim q \lesssim 2\pi/l, \\ (d/L)^{D_m}(ql/2\pi)^{-4}, & 2\pi/l \lesssim q. \end{cases} \tag{4.48}$$

Such a behaviour of the form factor can be obtained from a structure similar to the one in Fig. 4.10 in the basic units (disks) generate k other units (also, disks here) and with size scaled by the scaling factor $\beta < 1$. If we denote by r_0 the size of disk at $m = 0$, then at mth iteration $l = \beta^n r_0$. Thus, the last equations give [13]:

$$\frac{I_m^{(sf)}(q)}{V_0} = \begin{cases} \frac{1-\beta^{2(m+1)(3-D_m)}}{1-\beta^{2(3-D_m)}} & \text{at} \quad q = 2\pi/L, \\ \beta^4 + \beta^{6-D_m} \frac{1-\beta^{(2m+1)(3-D_m)}}{1-\beta^{2(3-D_m)}} & \text{at} \quad q = 2\pi/(\beta L), \\ \beta^8 + \beta^{10-D_m} + \beta^{2(6-D_m)} \frac{1-\beta^{(2m-1)(3-D_m)}}{1-\beta^{2(3-D_m)}} & \text{at} \quad q = 2\pi/(\beta^2 L), \end{cases} \tag{4.49}$$

In the limit of high number of iterations, the last equation can be simplified to:

$$\frac{I_m^{(sf)}(q)}{V_0} = \begin{cases} \frac{1}{1-\beta^{2(3-D_m)}} & \text{at} \quad q = 2\pi/L, \\ \beta^4 + \frac{\beta^{6-D_m}}{1-\beta^{2(3-D_m)}} & \text{at} \quad q = 2\pi/(\beta L), \\ \beta^8 + \beta^{10-D_m} + \frac{\beta^{2(6-D_m)}}{1-\beta^{2(3-D_m)}} & \text{at} \quad q = 2\pi/(\beta^2 L), \end{cases} \tag{4.50}$$

Since the scaling factor is $\beta < 1$ and the fractal dimension satisfies the condition $2 < D_m = D_s < 3$, one can further neglect the terms β^4 and $\beta^8 + \beta^{10-D_m}$ in Eq. (4.50), thus giving us [13]:

$$\frac{I_m^{(sf)}(2\pi/L)}{I_m^{(sf)}(2\pi/\beta L)} \simeq \frac{I_m^{(sf)}(2\pi/\beta L)}{I_m^{(sf)}(2\pi/\beta^2 L)} \simeq \beta^{D_s-6}. \tag{4.51}$$

A similar behavior can be obtained at arbitrarily values of the scattering vectors $q = 2\pi/(\beta^n L)$, with $n \leq m$. For 2D structures it can be shown, following a similar procedure that:

$$\frac{I_m^{(sf)}(2\pi/L)}{I_m^{(sf)}(2\pi/\beta L)} \simeq \frac{I_m^{(sf)}(2\pi/\beta L)}{I_m^{(sf)}(2\pi/\beta^2 L)} \simeq \beta^{D_s-4}. \tag{4.52}$$

Therefore, the surface fractal intensity given by Eq. (4.46) obeys approximately the power-law decay with the exponent $6 - D_s$ for 3D structures, and $4 - D_s$ for 2D structures.

4.3.4 Approximation of Independent Units

By denoting with $I_0(q) \equiv V_0^2 \langle |F_0(q)|^2 \rangle$ the intensity at zero-th iteration, and making the approximation $\langle |F_m^{(mf)}(q)|^2 \rangle \simeq k^{-n} \langle |F_0(q)|^2 \rangle$, Eq. (4.47) gives [13]:

$$I_m^{(sf)}(q) = \sum_{n=0}^{m} \beta^{n(6-D_m)} I_0 \left(\beta^n q \right), \tag{4.53}$$

which is called the approximation of incoherent amplitudes of the basic units, or simpler the approximation of independent units.

Let's recall that the Porod-law behaviour of scattering intensity at $m = 0$ can be written qualitatively as:

$$\frac{I_0(q)}{I_0(0)} \simeq \begin{cases} 1 & q \lesssim 2\pi/r_0, \\ q^{-4} & q \gtrsim 2\pi/r_0. \end{cases} \tag{4.54}$$

Then, when $q \lesssim 2\pi/r_0$ the term $I_0(q)$ in Eq. (4.53) dominates, since $\beta^{6-D_m} \ll 1$. When $q \simeq 2\pi/\beta r_0$ it is clear that its weight is reduced $1/\beta^4$ times, while the term $\beta^{6-D_m} I_0(\beta q)$ remains unchanged. As a consequence the term $\beta^{6-D_m} I_0(\beta q)$ dominates at $q \simeq 2\pi/\beta r_0$, if the condition $6 - D_s < 4$ is satisfied. Generally, one can see that for an arbitrarily iteration number, at the point $q \simeq 2\pi/\beta^{n-1} r_0$ the nth term in Eq. (4.47) dominates. Thus, when $q \to \beta q$, then $I(q) \to I(q)/\beta^{6-D_s}$, while the slope of the scattering curve is given by $\tau \equiv \log \left(1/\beta^{D_s-6} \right) / \log (1/\beta) = D_s - 6$.

Figure 4.11 shows numerical results for cases when the ratio of the distances between the scattering units and their size is $d/l = 1$, and respectively $d/l = 5$. The results clearly shows that by increasing the ratio d/l, the approximation of independent units given by Eq. (4.53) (red—dashes line) to the incoherent mass fractal amplitudes given by Eq. (4.47) is much better, and only the Porod regions of the mass fractals contribute to the total intensity of the surface fractal, which means that the approximation of independent units is applicable. This works only when the ratio d/l of distances between the basic units and their size is much bigger than one. This happens due to the fact that as d increase, the correlations between amplitudes decay very fast. Therefore, the surface fractal regime arise when the correlations inside a given mass fractal iteration or the influence of additional terms of mass-fractal intensities are negligible.

The contributions of each mass fractal iteration composing the surface fractal are shown in Fig. 4.11. Their corresponding intensities obeys the condition $\langle |M_0(0)|^2 \rangle > \langle |M_1(0)|^2 \rangle > \cdots > \langle |M_n(0)|^2 \rangle$ since the contribution of their respective volumes/areas to the total volume/area of the surface fractal decreases, and $\langle |F_n^{(mf)}(0)|^2 \rangle = 1$ [8, 13]. Recall that at nth iteration the corresponding volumes are $V_n = V_0 \beta^{n(3-D_m)}$. At $n = 0$ the scattering intensity dominates in the Guinier regime at $qL < 2\pi$ since it has the largest volume/area among as compared with other iterations. However, at $qL \gtrsim 2\pi$ it decays as $1/q^4$, which is much faster than the

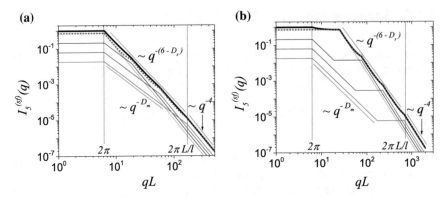

Fig. 4.11 Scattering intensity from a surface fractal at $m = 3$ as a function of the momentum transfer for fractal dimension $D_s = D_m = 2.45$, $L/l = 100$ and **a** $d/l = 1$; **b** $d/l = 4$ (see Ref. [13]), where L is the overall size of the fractal, l is the size of the units composing the fractal, and d is the distance between them. The solid (black) line shows the approximation of incoherent mass-fractal amplitudes (Eq. (4.53)). The dotted (red) line shows the approximation of incoherent amplitudes of the primary objects (Eq. (4.47)). The contribution of each mass fractal iteration is shown in various colors: $n = 0$ (green), $n = 1$ (blue), $n = 2$ (magenta), and $n = 3$ (dark yellow). The vertical lines indicate the limits of the fractal regions in the two cases

fractal power-law decay $1/q^{D_m}$ arising from the first iteration. Thus, the later term will dominate in this regime, up to $q \simeq 2\pi/(\beta L)$, after which the second iteration will dominate, and so on. In Fig. 4.11a we have $d/l \simeq 1$, the correlations between the composing basic units at each mass fractal iterations are important, and thus the approximation of incoherent amplitudes of each basic unit is not applicable. However, when $d/l \gg 1$ (Fig. 4.11b) only the Porod regions of each mass fractal contribute significantly to the total intensity, and therefore the approximation of independent units can be applied.

4.3.5   Application to SAS from Cantor Surface Fractals

We shall illustrate the above conclusions to two representative examples with $d/l \simeq 1$ and $d/l \gg 1$. For this purpose we consider a 2D CSF, which can be adapted so that both cases can be easily presented and discussed. The construction process of CSF has been illustrated in Sect. 4.3.2 and their models are shown in Fig. 4.10 for $d/l \simeq 1$ and respectively in Fig. 4.12 for $d/l \gg 1$. Here we emphasize that the single disk considered as the initiator at $n = 0$ has the radius r_0, the $k = 4$ disks at $n = 1$ have the radius $r_1 = \beta r_0$, the k^2 disks at $n = 2$ have the radius $r_2 = \beta^2 r_0$ and so on. Thus, the corresponding surface area of each individual mass fractal are πr_0^2 at $n = 0$, $k\pi r_1^2$ at $n = 2$, $k\pi r_2^2$ at $n = 2$ etc. It turns out that and the key difference between CMF and CSF is that while at a given iteration CMF consists of basic units of the same

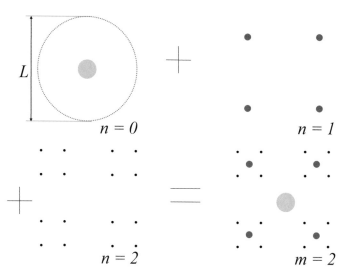

Fig. 4.12 Construction of the CSF at iteration $m = 2$ as a sum of CMF at iterations $n = 0$ (green—light gray), $n = 1$ (red—gray) and $n = 2$ (black—dark) for ratio $d/l \gg 1$ (see Ref. [8]). L is the overall size of the fractal

size, CSF have basic units of different sizes. The centers of the disks at $m = 1$ are shifted from the origin by:

$$a_j = L\{\pm (1 - \beta)/2, \pm (1 - \beta)/2\}, \tag{4.55}$$

with all possible combinations of the sign. The number of units at mth surface fractal iteration is:

$$N_m = \sum_{n=0}^{m} k^m = \left(k^{m+1} - 1\right)/(k - 1), \tag{4.56}$$

and the total surface area is:

$$S_m = S_0 \left(1 - k\beta^2\right)^{m+1}/\left(1 - k\beta^2\right), \tag{4.57}$$

where $S_0 = \pi r_0^2$.

Note that the surface fractal dimension of CSF coincides with the one of CMF given by Eq. (4.11), and is fixed at about 1.26. However, more general models such as those developed in Refs. [13] allow for arbitrarily values of the fractal dimension in the range $1 < D_s < 2$ by allowing the variation of the scaling factor, thus leading to what is known in the literature as generalized CSF (GCSF). In such cases a condition to avoid overlapping between different iterations shall be imposed. For the GSCF this is given by $r_0 \leq L (1 - 2\beta)/2$, and for our model an exact value is obtained by setting $\beta = 1/3$.

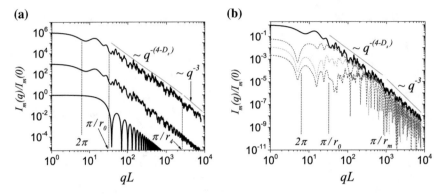

Fig. 4.13 SAS from 2D CSF (see Ref. [8]). **a** Iterations $m = 0, 2$ and 4, scaled up for clarity by a factor of 10^{3m}. The limits of the main regions are indicated by vertical dashed lines for $m = 4$. At $m = 2$ the upper fractal border region is at $\pi/r_2 \simeq 283$. **b** Iteration $m = 3$ (black—continuous line). Dashed lines represent SAS from CMF. From top to bottom: $n = 1$ (red), $n = 2$ (green), $n = 3$ (blue). Here, $D_s \simeq 1.26$ is the fractal dimension, and r_m denotes the radius of the disk at mth iteration

According to the model construction, the CMF at nth iteration consists of disks of radius $r_m = \beta^n r_0$ and with normalized scattering amplitude given by Eq. (4.17), with $r_0 = a/2$. Since the CSF is a sum of CMF at various iterations, the total scattering amplitude of CSF is given by the addition of amplitudes of each composing CMF, and therefore [13]:

$$F_m^{(sf)}(q) = \frac{1 - k\beta^2}{1 - \left(k\beta^2\right)^{m+1}} \sum_{n=0}^{m} \left(k\beta^2\right)^n F_n^{(mf)}(q),\tag{4.58}$$

with normalization $F_m^{(sf)} = 0$. Therefore, the scattering intensity of CSF is:

$$I_m^{(sf)}(q) = I_m^{(sf)}(0) \left\langle |F_m^{(sf)}(q)|^2 \right\rangle,\tag{4.59}$$

where $I_m^{(sf)}(0) = n|\Delta\rho|^2 S_m^2$, and $S_m = S_0 \left(k\beta^2\right)^m$. Figure 4.13a shows the corresponding SAS intensities for $m = 0, 2$ and 4. The main features of the curves are the presence of the three main regions: Guinier ($I(q) \propto q^0$), fractal ($I(q) \propto q^{D_s-4}$) and Porod ($I(q) \propto q^{-3}$) as in the case of scattering from CMF. However for CSF an additional intermediate regime ($I(q) \propto q^0/k^m$) occurs, and situated between the Guinier and fractal regimes. Below we shall discuss each region separately.

In the Guinier region, i.e. at $q \lesssim 2\pi/L$ one has only coherent scattering of the disks with zero phase difference. Therefore, at distances of the order of the overall size L of the fractal, the spatial correlations play an important role. In the intermediate region, i.e. at $2\pi/L \lesssim q \lesssim \pi/r_0$ we have an interference of the scattering amplitudes of CMF whose correlations decay with increasing q. If we denote by r_{nj} the typical distances between disks composing the mth, and respectively the jth iterations of CMF, then the correlations between amplitudes of different CMF iterations satisfy the relation $\left\langle F_n^{(\mathrm{mf})}(\boldsymbol{q}) F_j^{(\mathrm{mf})}(\boldsymbol{q}) \right\rangle \simeq 0$ for $n \neq j$ and $q \gtrsim 2\pi/r_{nj}$. Thus, by using Eqs. (4.58) and (4.59), we can write [13]:

$$I_m^{(\mathrm{sf})}(q)/I_m^{(\mathrm{sf})}(0) \equiv \left\langle |F_m^{(\mathrm{sf})}(\boldsymbol{q})|^2 \right\rangle \simeq \frac{\left(1 - k\beta^2\right)^2}{\left[1 - \left(k\beta^2\right)^{m+1}\right]^2} \sum_{n=0}^{m} \left(k\beta^2\right)^{2n} \left\langle |F_m^{(\mathrm{mf})}(\boldsymbol{q})|^2 \right\rangle,$$

(4.60)

with $k = 4$ for 2D CSF. The contribution of each CMF to the total scattering intensity of CSF is shown in Fig. 4.13b. It is clear from the figure that for a given mass-fractal iteration n, the spatial correlations are not important when q is of the order or higher than the end of the fractal region of the mass fractal, and therefore in this region one can use the approximation $\left\langle |F_m^{(\mathrm{mf})}(\boldsymbol{q})|^2 \right\rangle \simeq F_0^2\left(\beta^n q\right)/k^n$. By using this property in Eq. (4.60), we obtain [13]:

$$I_m^{(\mathrm{sf})}(q)/I_m^{(\mathrm{sf})}(0) \equiv \left\langle |F_m^{(\mathrm{sf})}(\boldsymbol{q})|^2 \right\rangle \simeq \frac{\left(1 - k\beta^2\right)^2}{\left[1 - \left(k\beta^2\right)^{m+1}\right]^2} \sum_{n=0}^{m} k^n \beta^{4n} F_0^2\left(\beta^n q r_0\right),$$

(4.61)

which describes the intensity when the correlations between the amplitudes of all disks composing the CSF are negligible. The upper border of the intermediate region ends where each disk stops to behave as a point-like object, and thus $q \lesssim 2\pi/r_0$. The intensities from each CMF in this region are proportional to their surface areas $S_0 \beta^{2n}$. Thus, the asymptotic value of the intensity of CSF in the intermediate region can be obtained by replacing $F_0 = 1$ in Eq. (4.61) and summing only the remaining terms. Such a replacement is motivated by the fact that in the intermediate region the correlations between disks amplitudes are very small but still the disks scatter as point-like objects. Thus, the asymptotic value becomes [13]:

$$I_m^{(\mathrm{sf})}(q)/I_m^{(\mathrm{sf})}(0) \equiv \left\langle |F_m^{(\mathrm{sf})}(\boldsymbol{q})|^2 \right\rangle \simeq \frac{\left(1 - k\beta^2\right)^2}{\left[1 - \left(k\beta^2\right)^{m+1}\right]^2} \frac{1 - \left(k\beta^4\right)^{m+1}}{1 - k\beta^4}.$$

(4.62)

This region can be also considered as a Guinier region corresponding to the CSF consisting of spatially uncorrelated objects, and whose main contribution comes from CMF at $n = 1$.

In the fractal region, the contributions to the total scattering arise from the central disk of surface area S_0 and intensity $I_0(q) \equiv n|\Delta\rho|^2 S_0^2 F_0^2(qr_0)$, from the $k = 4$ disks with $S_1 = S_0 k\beta^2$ and $I_1(q) = k\beta^4 I_0(\beta q)$ and so on. By analogy with the derivations performed in the previous section, we can find that the scattering exponent is $D_s - 6$ in the surface fractal region. Since we need to add the intensities but not the amplitudes. the ratio d/l shall be high enough so that the surface fractal region can be obtained as an incoherent diffraction of all disks. The upper border of the fractal region, i.e. when $q \lesssim \pi/\left(\beta^m r_0\right)$, is limited by the size of the smallest disks inside

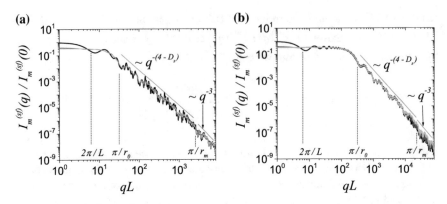

Fig. 4.14 SAS from 2D CSF with fractal dimension $D_s \simeq 1.26$ at $m = 4$ for **a** $d/l = 10$, and **b** $d/l = 100$ (see Ref. [8]), where d is the distance between units (disks) composing the fractal, and l is their size.. Black—total intensity (Eq. (4.59)), Green—neglecting correlations between mass fractal amplitudes (Eq. (4.60)), Red—approximation of independent units (Eq. (4.61)). The coincidence in the fractal region, between the three curves becomes better with increasing the ratio d/l. L is the overall size of the fractal

the fractal. In the Porod region, i.e. at $q \gtrsim \pi/(\beta^m r_0)$ the total scattering intensity decays as q^{-3} and is given by the intensity of the single disk at $m = 0$.

Figure 4.14a shows that the total scattering of CSF in the fractal region arise from the non-coherent sum of the disks. Also, for the numerical values of the control parameters used, i.e. $d/l = 10$, the agreement between the total intensity (Eq. (4.59)), the approximation neglecting the correlations between mass-fractal amplitudes (Eq. (4.60)) and the approximation of the independent units (Eq. (4.61)) is not quite satisfactory. However, by increasing the ratio to $d/l = 100$, the agreement between the three curves becomes much better, as shown in Fig. 4.14b. This is in accordance with the discussion form the previous section. Note that starting with $qL \gtrsim 10^4$, due to numerical errors in computation of scattering intensities at such large values of momentum transfer, the approximation of independent units decays slightly slower than expected. Furthermore, a plot of $I(q)q^{6-D_s}$ versus q reveal also a log-periodicity of the curve with the period $\log 1/\beta$, and with the number of minima coinciding with the iteration number, as in the case of CMF. Although the borders of this log-periodicity are given by the limits of the fractal range, for CSF their nature is different as compared with CMF. While in the later case, they arise from the self-similarity property of the distances between the structural units, in the former one it arises from their size.

Figure 4.15 (left part) shows a simulation the structure of the 2D CSF at $m = 2$ using Monte Carlo methods, for generating the $p(r)$ function. The algorithm for generating the points is similar to the one employed in generating the 2D CMF shown in Fig. 4.6 (left part). For the CSF at $m = 2$ we have used a total number of 3×10^4 points. Those found inside the region delimited by the CSF (red points) are kept (about 8×10^3 points), while the others (gray) are discarded. Therefore,

 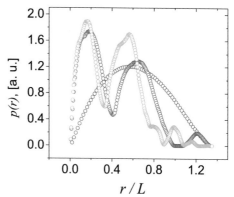

Fig. 4.15 Left part: Approximating the second iteration ($m = 2$) of 2D CSF set with randomly generated points (red). Right part: the pair distance distribution functions for $m = 0$ (blue), $m = 1$ (red) and $m = 2$ (green) of the 2D CSF simulated using randomly generated points. L is the overall size of the fractal

the total number of calculated distances is about 35.28×10^6. The corresponding $p(r)$ function of CSF at $m = 0$ (blue), $m = 1$ (red) and $m = 2$ (green) are shown in Fig. 4.15 (right part). For $m = 0$ we have the $p(r)$ function corresponding to s single disk, as expected. For $m = 1$ and $m = 2$ we have a succession of maxima and minima which reflect the most and respectively the least probable distances within the fractal, at a given ratio r/L. However, a clear periodicity as in the case of CMF is hardly observable, since we have now a distribution of disks of different sizes. At $m = 1$ and $m = 2$, only the positions of first two maxima can be used to recover the fractal scaling factor. If we know a priori that the fractal has the same scaling factor at each iteration then, we can use this value for the whole number of iterations. However, if the value of the scaling factor is iteration-dependent (as we shall see in a later section, the case of fat fractals), then only the scaling factor at first iteration is obvious. Additional information could be extracted, in principle, from the succession of maxima and minima at high r/L values.

4.3.6 Case Study: Koch Snowflakes

As an application to SAS from surface fractals, we consider in the following a planar silver Koch surface fractal fabricated by electron beam lithography directly onto a 50 nm thick silicone TEM grids, where poly(methylmethacrylate) has been deposited by spin coating at 6000 rpm [14]. In a second step, the structure was baked for 2 min at 144 K while the development has been performed for about 70 s in a isopropyl alcohol isobutyl ketone solution at a ratio 3:1, followed by isopropyl alcohol rising and drying with N_2. The silver has been deposited in an electron beam evaporator system and the lift-off was performed by soaking the sample in acetone and again rising with isopropyl alcohol. The as-obtained structure is a Koch snowflake (KS) at first iteration with the overall size of 2 μm, which shown in Fig. 4.16 (Left part).

In building the structural model, we start from a single triangle of edge a and area $S^T = \sqrt{3}a^2/4$, which represents also a mass fractal at iteration $n = 0$. Then, the first iteration of KS is obtained by dividing each edge into three equal segments and adding an outward oriented equilateral triangle with the base coinciding with the central segment. As a result, three new triangles of edge length $a/3$ are added to the initial triangle of edge a. Higher iterations are obtained by repeating the same procedure to each new line segment. At mth iteration the KS consists of 3×4^m number of equal edges of length $a_m = a/3^m$. Thus, the fractal dimension of the perimeter is $D_s = \lim_{m\to\infty} \log(3 \times 4^m)/\log(a/a_m) \simeq 1.26$. Within this model, the nth mass fractal iteration consists of $N_n = 3 \times 4^{n-1}$, $n = 1, 2, \cdots, m$ triangles of edge size $a_n = a/3^n$, area $N_n S^T/3^{2n}$, and with a mass fractal dimension D_m which coincide with D_s.

For convenience, we consider that the zero-th iteration of KS consists of the sum of the single triangle of edge a and the three triangles of edge $a/3$. This structure is known also in the literature as the Star of David. Therefore, at an arbitrarily iteration m, the total area of KS is:

$$S_m = S^T + \sum_{n=1}^{m+1} \frac{S^T}{3^{2n}} N_m = \frac{4S^T}{5}\left(2 - \frac{1}{3}\frac{4^m}{9^m}\right). \tag{4.63}$$

Figure 4.16 (Right part) shows the KS at second iteration, where each composing mass fractal has been emphasized by different color.

In order to obtain the scattering amplitude, we make use of the scattering properties presented in Sect. 4.2.3, which allow us to write down the following recurrence formula [8]:

$$A_m(\mathbf{q}) = 6G_2(\mathbf{q})\left[\beta^2 A_{m-1}(\beta\mathbf{q}) - 6G_1(\beta\mathbf{q})\beta^4 A_{m-2}(\beta^2\mathbf{q})\right]$$
$$+ \beta^2 A_{m-1}(\beta\mathbf{q})\left[1 + 6G_1(\mathbf{q})\right], \tag{4.64}$$

where $G_1(\mathbf{q}) = (1/6)\sum_{j=0}^5 \exp(-i\mathbf{q} \cdot \mathbf{c}_j)$, $G_2(\mathbf{q}) = (1/6)\sum_{j=0}^5 \exp(-i\mathbf{q} \cdot \mathbf{b}_j)$ are the generative functions, and $\beta = 1/3$. Here, the position vectors $\mathbf{c}_j$ and $\mathbf{b}_j$ are given by $\mathbf{c}_j = (2a)/9\{\cos(\pi(j+1/2)/3), \sin(\pi(j+1/2)/3)\}$ and respectively by $\mathbf{b}_j = (2a)/\left(3\sqrt{3}\right)\{\cos(\pi j/3), \sin(\pi j/3)\}$. This form of the scattering amplitude allows us to obtain the non-normalized scattering amplitude of KS at arbitrarily iteration number, knowing the amplitudes at $m = 0$ and $m = 1$, where the form factor of a triangle is given by Eq. (3.81). Thus, the scattering intensity can be written as $I_m(q) \propto \langle|A_m(\mathbf{q})|^2\rangle$.

Equation (4.64) allows also to obtain an analytic expression for the radius of gyration R. We present here only the result in the following form [8]:

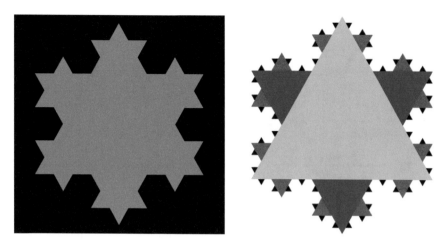

Fig. 4.16 Left part: the KS with an overall size of 2 μm by electron beam lithography (see Ref. [14]). Right part: a model of KS used to calculate the SAS intensities. All triangles of a given size have the same color, and represent a mass fractal iteration. Green—$n = 0$, blue—$n = 1$, red—$n = 2$, black—$n = 3$. In this representation, the obtained KS on the left part corresponds to the model consisting from green, blue and red triangles, from the right part

$$R_m^2 = \frac{351 R_{m-1}^2 S_{m-1} - 12 R_{m-2}^2 S_{m-2} + 32a^2 (9 S_{m-1} - 2 S_{m-2})}{2187 S_m}, \qquad (4.65)$$

which for $m = 1$ gives $R_1^2 = 223a^2/1944$. In the ideal case, i.e. for $m \to \infty$, we obtain the radius of gyration of KS as [8]:

$$R_{KS} = 2a/\sqrt{33}. \qquad (4.66)$$

The corresponding monodisperse and polydisperse scattering intensities at $m = 1$ and $a = 2$ μm are shown in Fig. 4.17a. For the polydisperse curve, the log-normal distribution function of the sizes given by Eq. (3.69) with the relative variance $\sigma_r = 0.2$ has been used. The general scattering properties are similar to those of 2D CSF discussed in the previous section. The upper edge of the Guinier region is at $q \simeq 2\pi/a = 3.14 \times 10^4$ Å^{-1}, while the fractal range is found between $q \simeq 2\pi/(a\beta) = 9.42 \times 10^4$ Å^{-1} and $q \simeq 2\pi/(a\beta^2) = 28.3 \times 10^4$ Å^{-1}. Within this range, the expected value of the fractal dimension is $D_s = 1.26$ and thus the scattering exponent is $\tau = 2.74$. However, since we have only one iteration, the length of the fractal region is quite short for a reliable determination of the fractal dimension. To solve this issue, samples with at least one more iterations are required.

Figure 4.17b shows the $p(r)$ function of the KS (red—discrete stars), and for comparison, the $p(r)$ of a disk centered on KS and circumscribed to it, i.e. having the radius $a\sqrt{3}/3$. The results show that the $p(r)$ of KS is very similar to the one corresponding to the disk, thus reflecting the high degree of symmetry. However, there are some difference which need to be emphasized. First, in the case of KS, in

(a)   **(b)**

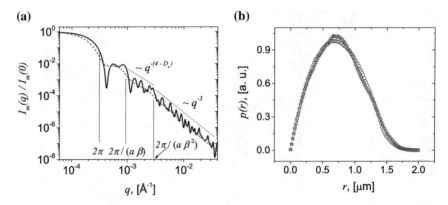

Fig. 4.17 **a** SAS intensity from KS surface fractal at $m = 1$ (see Ref. [8]). Black (continuous line)—monodisperse form factor, red (dashed line)—polydisperse form factor with the distribution function given by Eq. (3.69) with relative variance $\sigma_r = 0.2$ **b** $p(r)$ function of the same KS (red—discrete stars). For comparison, the $p(r)$ of a disk (blue—discrete circles) centered and circumscribed to KS has been included. Here, $a \simeq 2\mu$m is the overall size of KS, $\beta = 1/3$ is the scaling factor, and $D_s \simeq 1.26$ is the fractal dimension

the range $0.25\ \mu\text{m} \lesssim r \lesssim 1.25\ \mu\text{m}$ the distribution of distances in the $p(r)$ function is slightly narrower. Second, at $r \simeq 1.25\ \mu\text{m}$ a kink arise, followed by a slower decay as compared to the $p(r)$ of the disk, while at $r \gtrsim 1.7\ \mu\text{m}$ the two functions coincide completely. These observations reflect the rougher structure of the perimeter or KS.

Note that for the model shown in Fig. 4.16 (Right part) the ratio between the distances between triangles composing the fractal and their size is of the order of unity, and therefore, as discussed in the previous section, the approximation of neglecting the correlations between mass fractal amplitudes (Eq. (4.60)), and the approximation of independent units (Eq. (4.61)) will not work.

4.4   Heterogeneous Fractals

4.4.1   Variation of the Number of Basic Units with Iteration Number

Many experimental SAS data [15–17] are characterized by a succession of power-law decays with different exponents. In the case of a succession of two power-law regimes with exponents τ_1 and τ_2, we distinguish two cases. If $\tau_1 < \tau_2$ we have a convex decay, while for $\tau_1 > \tau_2$ we have a concave decay. Recently, theoretical models have been proposed for both cases. For 'convex' data the model involves a three-phase structure in which homogeneous structures with SLD ρ_1 and ρ_2 are immersed in a homogeneous matrix/solution with SLD ρ_0. In the case of a 'concave' succession

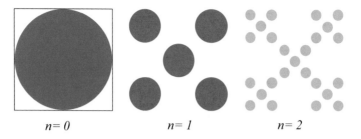

Fig. 4.18 First three iterations $n = 0$ (initiator), $n = 1$ (generator) and $n = 2$ of the 2D Vicsek fractal (see Ref. [4])

one starts from a two-phase system in which the homogeneous structures with SLD ρ_1 are fractals whose scaling factors vary with the iteration number. Since three-phase systems can be reduced to two-phase systems by subtracting the "background" density, here we also present and discuss results describing two-phase systems. The models are based on deterministic mass fractals, and we show that the behavior given by a succession of 'convex'/'concave' SAS data, and more generally, by any combination of them, can be described in terms of the relative positions and the number of structural subunits of the fractal, and which vary with the iteration number.

For our purpose we consider in the following a 2D Vicsek fractal (VF) of overall size a, which is very similar to the CMF but with the difference that at first iteration, in addition to the four corner disks, the middle one is also kept, as shown in Fig. 4.18. Thus, at nth iteration one obtains a structure with $k_n = 5^n$ disks (all of equal size), fractal dimension $D_{VF} \simeq 1.46$, and the generative function given by:

$$G_n(\boldsymbol{q}) = \frac{1}{5} \left(\cos\left(q_x a\beta^n\right) \cos\left(q_y a\beta^n\right) + 1 \right). \tag{4.67}$$

Thus, in order to observe a succession of power-law decays with different exponents, we proceed as follows: for the first ith iterations we consider the structure of a single fractal (either CMF or VF), while for the next jth iterations we consider the structure of the other fractal. Then, depending on the order in which we consider the fractals, we shall expect a succession of power-law decays either of the form $q^{-D_{CMF}} \to q^{-D_{VF}}$ or $q^{-D_{VF}} \to q^{-D_{CMF}}$, where $D_{CMF} \simeq 1.26$ is the fractal dimension of the 2D CMF, given by Eq. (4.11).

Figure 4.19a and b shows the monodisperse and polydisperse scattering structure factors for the above two cases, where CMF and VF are considered for three consecutive iterations, i.e. $n = 1, 2, 3$ and $n = 4, 5, 6$. For the polydisperse curve, the log-normal distribution given by Eq. (3.69) with the relative variance $\sigma_r = 0.2$ has been used. The results show that when VF is used for the first three iterations followed by CMF at next three iterations we obtain a 'convex' succession of the type $q^{-1.46} \to q^{-1.26}$, while when CMF is used for the first three iterations followed by VF at next three iterations we have a 'concave' succession $q^{-1.26} \to q^{-1.46}$. The

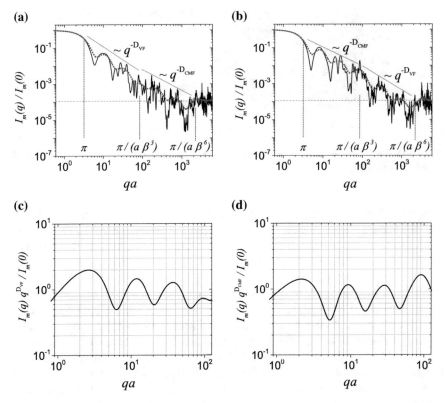

Fig. 4.19 Monodisperse (black—continuous line) and polydisperse (red—discrete line) structure factors from a hierarchical system, illustrating a 'convex' and respectively a 'concave' succession of power-law decays (see Ref. [18, 19]). **a** VF with fractal dimension $D_{VF} \simeq 1.46$ at iterations $m = 1, 2, 3$ followed by CMF with fractal dimension $D_{CMF} \simeq 1.26$ at $m = 4, 5, 6$. **b** CMF at iterations $m = 1, 2, 3$ followed by VF at $m = 4, 5, 6$. Vertical dotted lines—limits of the fractal regions. Horizontal dashed line—asymptotic values of the structure factor. **c** The quantity $I_m(q)q^{D_{CMF}}$ versus q. **d** The quantity $I_m(q)q^{D_{VF}}$ versus q. Here, a is the overall size of the fractal, and $\beta = 1/3$ is the scaling factor

range for which each individual power-law regime holds is delimited by vertical dotted lines in Fig. 4.19a and b. In both cases, the asymptotic value of the structure factor is $S(q) \rightarrow (1/4^i)(1/5^j) = (1/4^j)(1/5^i)$, since $i = j = 3$, which reflects the number of structural units in the system.

The log-periodicity of the polydisperse structure factors is more clearly seen by representing them in the form $I(q)q^{D_{CMF}}$ versus q, and $I(q)q^{D_{VF}}$ versus q, as shown in Fig. 4.19c and d. Thus, the number of iterations and the scaling factor for each individual power-law decay can be also determined for hierarchical systems. Other quantities of interest, such as the radius of gyration, the overall shape or the pddf can be obtained in a similar manner as for the simple CMF, through Guinier/Porod plots and so on (see Fig. 4.8).

4.4.2 Variation of the Scaling Factors with Iteration Number

An important class of heterogeneous fractals can be constructed when instead of changing the number of basic units forming the fractal, we change the value of the scaling factor. In particular, if their numerical values decrease with iteration number one obtains the so-called *fat fractals*, i.e. fractals with positive Lebesgue measure. This leads to appearance of a succession of power law decays in SAS intensity, with decreasing values of the scattering exponents [18–20]. In the following we shall present a general model based on 2D CMF which can be used to systems showing a broad range of fractal dimensions.

Let's consider, as before, an initial square of edge length a at zero-th iteration in which is inscribed a disk of radius $r_0 = a/2$, with $r_0 \leq a/2$, and whose centers coincide with the origin of a Cartesian coordinate system. The edges of the square are parallel to the system's axes. Therefore, the coordinates of a point $P(x, y)$ inside the square satisfies the conditions $-a/2 \leq x \leq a/2$ and $-a/2 \leq y \leq a/2$. First iteration (i.e. $m = 1$) is the same as in the case of the regular CMF, and is obtained by dividing the initial square into 4 smaller squares of edge length $\beta_s^{(1)}a$ with positions given by Eq. (4.55), and with $0 < \beta_s^{(1)} < 1/2$. The quantity placed between $(\cdots)$ of the exponent of β_s shall be interpreted as an index quantifying the iteration number, and not as a power. Thus, the position vectors of the 4 squares, can be written as:

$$a_j = \{\pm\beta_t^{(1)}a, \pm\beta_t^{(1)}a\}, \tag{4.68}$$

with all combinations of the signs and $\beta_t^{(1)} = \left(1 - \beta_s^{(1)}\right)/2$.

This procedure is repeated for the next p iterations, keeping constant the scaling factor at $\beta_s^{(1)}$. Then, for next q iterations (i.e. in the range p, $p + q$) we change the scaling factor to $\beta_s^{(2)}$, satisfying the condition $\beta_s^{(2)} > \beta_s^{(1)}$. For the next r iterations (i.e. in the range $p + q$, $p + q + r$) we set the scaling factor to $\beta_s^{(3)}$, with $\beta_s^{(3)} > \beta_s^{(2)}$ and so on (see Fig. 4.20). In the high limit of iterations we obtain a fat fractal whose basic units have still positive area. The regular CMF studied before is obtained by choosing $\beta_s^{(1)} = \beta_s^{(2)} = \beta_s^{(3)} = \cdots = 1/3$.

Since the succession of two power-law decays can be either direct, i.e $q^{-D_1} \to q^{-D_2}$, with $D_1 < D_2$, or indirectly, i.e. $q^{-D_1} \to q^0 \to q^{-D_2}$, we shall analyze here both situations. For this purpose we introduce the dimensionless quantity:

$$f \equiv a/r_0, \tag{4.69}$$

which quantifies the length of the intermediate region, where $I(q) \propto q^0$. Note, that a similar situation occurred also in scattering from CSF and where we have used the ratio d/l to describe the relative size d of the basic units compared to the distance l between them.

The difference between the corresponding structures when $f = 2$ and $f = 4$ is illustrated in Fig. 4.20 (Upper row) and respectively (Lower row), where the construction process of 2D fat CMF (FCMF) is shown at first $m = 3$ iterations. For

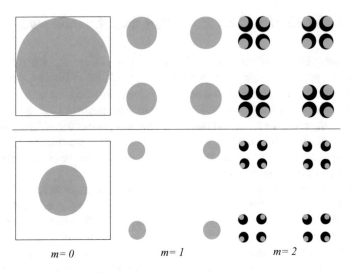

Fig. 4.20 A comparison between regular and fat fractals at iterations $m = 0$, 1 and 2, where at $m = 0$ the basic shape is a disk of radius $r_0 = a/2$ and the fractal size L (see Ref. [20]). In the model, at $m = 1$ the fractal has the scaling factor $\beta_s^{(1)}$, and at $m = 2$, the scaling factor is $\beta_s^{(2)}$, with $\beta_s^{(2)} > \beta_s^{(1)}$. Upper row: $f \equiv L/r_0 = 2$. Lower row: $f \equiv L/r_0 = 4$. In both cases, at m = 1 regular and fat fractals coincide due to equal scaling factors. However, starting with $m = 2$, the fat fractal has a bigger scaling factor and thus the corresponding disks have a larger diameter (black disks) than in the case of regular fractal (light blue—light gray disks)

an optimal visualization, in both cases one considers that $\beta_s^{(1)}$ is kept constant at $m = 0$ and $m = 1$, $\beta_s^{(2)}$ is kept constant at $m = 2$ and $m = 3$ and so on. The same figure illustrates also the situation when $\beta_s^{(1)} = \beta_s^{(2)} = 1/3$ in order to emphasize the differences between fat and regular fractals.

Generally, we can illustrate the dependence of the scaling factor with the iteration number by writing [18]:

$$\beta_s^{(m)} = \frac{1 - c\,(1/3)^{p_m}}{2}, \tag{4.70}$$

where $0 < c \leq 1$. The exponent is defined as:

$$p_m = \lfloor \frac{m-1}{v} \rfloor, \tag{4.71}$$

where the symbol $\lfloor \cdot \rfloor$ indicates the floor function (i.e. the greatest integer less than or equal to $\cdot$), and v gives the number of iterations or which the scaling factor is kept constant.

It can be easily shown that in the limit of large number of iterations, the fractal dimension of FCMF is 2. By denoting with a_1 the area removed at iteration $m = 1$, with a_2 the relative area removed at $m = 2$ and so on, we find that the total area (i.e. Lebesgue measure in 2D) remaining at mth iteration is $\prod_{i=1}^{m} (1 - a_i) > 0$ if

$\sum_{i=1}^{m} a_i < \infty$, and thus the model fulfills the defining properties of fat fractals (cite refs 29, 30 from EPJB).

Since the values of the scaling factors change with the iteration number, the generative function of the FCMF can be written as [18]:

$$G_m(\boldsymbol{q}) = \cos\left(q_x a \beta_{\mathrm{t}}^{(m)}/2\right) \cos\left(q_y a \beta_{\mathrm{t}}^{(m)}/2\right), \tag{4.72}$$

where $\beta_{\mathrm{t}}^{(m)} = \left(1 - \beta_{\mathrm{s}}^{(m)}\right)/2$, and $\beta_{\mathrm{s}}^{(m)}$ is given by Eq. (4.70). Thus, the FCMF form factor can be written as [18]:

$$F_m^{\mathrm{FCMF}} = F_0\left(q \prod_{i=1}^{m} \beta_{\mathrm{s}}^{(i)}\right) \prod_{i=1}^{m} G_i(q u_i), \tag{4.73}$$

where $u_i \equiv a \beta_{\mathrm{s}}^{(i)} \prod_{j=1}^{i=1} \beta_{\mathrm{s}}^{(j)}$, and $F_0(\cdot)$ is the form factor of the basic unit. In the case of only two structural levels, i.e. when we have $\beta_{\mathrm{s}}^{(1)} = \beta_{\mathrm{s}}^{(2)} = \cdots = \beta_{\mathrm{s}}^{(p)}$, and $\beta_{\mathrm{s}}^{(p+1)} = \beta_{\mathrm{s}}^{(p+2)} = \cdots = \beta_{\mathrm{s}}^{(r)}$, with $1 \le p \le i$, $p+1 \le r \le m-i$ and $i = 1, 2, \cdots$, the fractal size a is replaced by a/f, when $f \neq 2$.

The numerical results for monodisperse and polydisperse scattering intensity of FCMF for $m = 6$, $i = 3$, $\beta_{\mathrm{s}}^{(1)} = \beta_{\mathrm{s}}^{(2)} = \beta_{\mathrm{s}}^{(3)} = 0.38$ (i.e. $D^{(1)} \simeq 1.43$), $\beta_{\mathrm{s}}^{(4)} = \beta_{\mathrm{s}}^{(5)} = \beta_{\mathrm{s}}^{(6)} = 0.22$ (i.e. $D^{(2)} \simeq 0.91$) are shown in Fig. 4.21a and respectively b. As before, the size distribution function is a log-normal one, with the relative variance $\sigma_r = 0.2$.

The common feature in both cases, i.e. when $f = 2$ and $f = 20$ is the presence of a Guinier region at $qa \lesssim \pi$, a succession of two power-law regimes at $2\pi \lesssim qa \lesssim 2\pi/\left(a\beta_{\mathrm{s}}^{(1)^3}\beta_{\mathrm{s}}^{(2)^3}\right)$, and respectively at $\pi \lesssim qa \lesssim 2\pi/\left(fa\beta_{\mathrm{s}}^{(1)^3}\beta_{\mathrm{s}}^{(2)^3}\right)$,

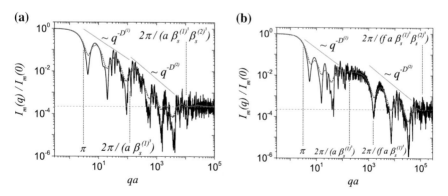

Fig. 4.21 Monodisperse (black—continuous line) and polydisperse (red—dotted line) structure factors of FCMF with scaling factors $\beta_{\mathrm{s}}^{(1)} = \beta_{\mathrm{s}}^{(2)} = \beta_{\mathrm{s}}^{(3)} = 0.38$ and $\beta_{\mathrm{s}}^{(4)} = \beta_{\mathrm{s}}^{(5)} = \beta_{\mathrm{s}}^{(6)} = 0.22$ at: **a** $f = 2$, **b** $f = 20$ (see Ref. [20]). The corresponding fractal dimensions of the two structural levels are $D^{(1)} \simeq 1.43$ and $D^{(1)} \simeq 0.91$. Vertical lines indicate the limits of the fractal regions. Horizontal line indicate the asymptotic value of the structure factor. Here, a is the overall size of the fractal

with each succession being followed by asymptotic regions with values approximating $1/k_m$.

It is known that in the Guinier region the scattering intensity can be approximated by Eq. (3.56). Thus, by expanding the form factor in Eq. (4.73) in power series in qa and substituting the result into Eq. (4.21) we obtain an analytic expression of the radius of gyration, as [18]:

$$R_g^{(m)} = \sqrt{R_{g0}^2 \prod_{i=1}^{m} \left(\beta_s^{(i)}\right)^2 + 3a^2 \sum_{i=1}^{m} \left(\beta_t^{(i)}\right)^2 \prod_{k=1}^{i-1} \left(\beta_s^{(k)}\right)^2}, \qquad (4.74)$$

where R_{g0} is the radius of gyration of the basic unit. Note, that when all the scaling factors are equal, Eq. (4.74) reduces to the well known expression of radius of gyration of CMF (see also Eq. (4.27)).

In the fractal region the scattering exponents coincide with the fractal dimension at each structural level. Between the two fractal regions, the intermediate Guinier region arise due to the high values of the ratio L/r_0. Note that in order to observe a power-law decay in either of the two structural levels, one needs to consider at least two consecutive iterations so that at least two minima are preserved.

The model can be easily generalized to 3D case by adding an extra cosine term of the type $\cos\left(q_z a \beta_t^{(m)}/2\right)$ in Eq. (4.72), to arbitrarily fractal dimensions by changing the values of the scaling factors, to fractal regions of various lengths by changing the number of iteration at a given structural level, or to other asymptotic values in the high q range by changing the expression of the generative function.

4.5 Multifractals

We construct a Vicsek-like [21] multifractal model by starting with a square of edge size a in which we inscribe a disk of radius r_0, with $0 < r_0 < a/2$ such as that their centers coincide (see Refs. [22, 23]). This disk is the zero-order iteration (i.e. $m = 0$). As before, We choose a Cartesian system of coordinates with the origin coinciding with the center of the square and disk, and with axes parallel to the square's edges. The first iteration ($m = 1$) is obtained by replacing the initial disk with $k_1 = 4$ smaller disks of radii $r_1 = \beta_{s1}a$ situated in the corners of the square, and with $k_2 = 1$ disks of radius $r_2 = \beta_{s2}a$ situated in the center. Thus, the fractal is build with the help of two scaling factors β_{s1} and β_{s2}. By performing a similar operation on each of the $k_1 + k_2$ disks one obtains the second fractal iteration ($m = 2$). For an arbitrarily number of fractal iterations m, the total number of disks composing the fractal is:

$$N_m = (k_1 + k_2)^m . \qquad (4.75)$$

Fig. 4.22 Second iteration ($m = 2$) of the multifractal models (see Refs. [22, 23]). Left part: $\beta_{s1} = 0.12$, $\beta_{s2} = 0.76$, $D^{I} \simeq 1.24$ (Model MFI). Middle: $\beta_{s1} = 0.24$, $\beta_{s2} = 0.52$, $D^{II} \simeq 1.35$ (Model MFII). Right part: $\beta_{s1} = 0.32$, $\beta_{s2} = 0.36$, $D^{III} \simeq 1.44$ (Model MFIII). Black—disks occurring at $m = 1$, orange (light grey)—disks occurring at $m = 2$

Therefore, by using Eq. (2.11) the box counting fractal dimension of the multifractal can be calculated according to [21]:

$$\sum_{i=1}^{2} k_i \beta_{si}^{D} = 1. \tag{4.76}$$

When the two scaling factors are equal to 1/3, this equation recovers the well-known fractal dimension for Vicsek fractal [21].

In Fig. 4.22 we show the second iteration of the multifractal model at various values of the scaling factors β_{s1} and β_{s2}. Black (dark) represents the disks which occur at first iteration, and orange (light grey) denotes the number of disks occurring at second iteration. The left part of Fig. 4.24 (model MFI) shows that for $\beta_{s1} = 0.12$ and $\beta_{s2} = 0.76$, a more heterogeneous structure is obtained as compared with the middle part i.e. for $\beta_{s1} = 0.24$ and $\beta_{s2} = 0.52$ (model MFII), or with the model MFIII, where $\beta_{s1} = 0.32$ and $\beta_{s2} = 0.36$. Note that for model MFIII the fractal consists of very closely sized disks, and it resembles the single scale Vicsek fractal, as expected. These differences in the heterogeneity are distinguishable in the multifractal spectra, pddf and SAS intensity, fact that will be emphasized below.

The corresponding dimension spectra D_q given by Eq. (2.36) for the three models MFI, MFII and MFIII are shown in Fig. 4.23a for $-25 < q < 25$. Here, q is an index and shall not be confused with the wavevector q appearing in figures showing scattering intensities. The standard terminology in literature is q in both cases. One can see that the dimension spectrum for model MFI (black curve) covers a broad range of values, with $0.32 \lesssim D_q \lesssim 1.74$. This behaviour arise due to the high degree of heterogeneity, with the densest regions having the fractal dimension $\simeq 1.73$, while the most rarefied ones have 0.32. The dimension spectrum for model MFII (red curve) covers the much tighter range between $1.02 \lesssim D_q \lesssim 1.5$. For this model, there still can be observed pronounced differences between regions with high and low densities. However, the spectrum of model MFIII (green curve) is almost a horizontal line, as expected, since the two scaling factors have close values. This leads to an almost homogeneous structure, with a fractal dimension of $D_q \simeq 1.44$.

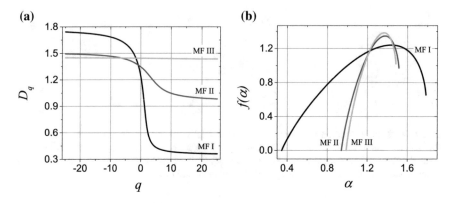

Fig. 4.23 Multifractal spectra at various scaling factors β_{s1} and β_{s2} for the models MFI ($\beta_{s1} = 0.12$, $\beta_{s2} = 0.76$), MFII ($\beta_{s1} = 0.24$, $\beta_{s2} = 0.52$) and MFIII ($\beta_{s1} = 0.32$, $\beta_{s2} = 0.36$; see Ref. [23]). **a** Generalized dimensions, where q is here only an index, and shall not be confused with the wavevector appearing in scattering curves. **b** Singularity spectra

Note that the box-counting dimensions can be determined using the intersection of the fractal dimension spectrum with $q = 0$ axis, and thus: $D_0 \simeq 1.24$ (for model MFI), $D_0 \simeq 1.35$ (for model MFII) and $D_0 \simeq 1.44$ (for model MFIII) [23]).

Figure 4.23b shows the singularity spectra $f(\alpha)$ versus α calculated by a box counting of measures using Eq. (2.41) for the three models. This allows to obtain the profile of $f(\alpha)$ through the implicit parameter q without the need to perform a numerical Legendre transform. The general properties of these spectra is that they behave as a single-hump functions with negative curvature. The width of the spectra is highest when $\beta_{s1} \ll \beta_{s2}$ (model MFI), and the position of maxima α_{max} are shifted to the left when β_{s1} becomes closer to β_{s2}. Finally, when $\beta_{s1} = \beta_{s2}$ it is known that the singularity spectrum is reduced to s single point. This behaviour reflects the heterogeneity of the structure, with the most heterogeneous one corresponding to model MFI. At maximum point $\alpha = \alpha_{max}$, the singularity spectra satisfy the conditions $f(\alpha_{max}) = D_0$ and $f'(\alpha_{max}) = 0$. In addition, the curve $f(\alpha)$ is tangent to the line $f = \alpha$ (not shown here).

Figure 4.24 shows the real space characteristics for the models MFI, MFII and MFIII at fractal iteration number $m = 4$, by calculating the coefficients C_p in Eq. (4.23), using a simple combinatorial analysis. They are presented on a double logarithmic scale, where we can clearly observe that the general feature is the presence of distance-groups.

For $\beta_{s1} = 0.12$ and $\beta_{s2} = 0.76$ the periodicity of the pair of distances is clearly observable in Fig. 4.24a, where each of the main distance-groups have a maximum of $C_p \simeq 400$ distances and they are separated by gaps at $r/l_0 \simeq 3.81 \times 10^{-3}$, 4.58×10^{-2} and respectively at 5.51×10^{-1}, indicating the absence of the corresponding distances inside the fractal. The position of these groups is well described as having the periodicity $\log_{10}(1/\beta_{s1})$, and thus, they are related to the scaling factor $\beta_{s1} = 0.12$. Inside each main group, other much less pronounced smaller gaps can be

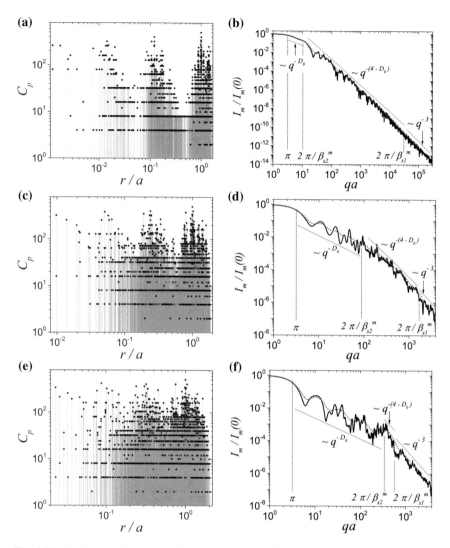

Fig. 4.24 Coefficients C_p given by Eq. (4.23) for the pddf (**a**), (**c**), (**e**), and the monodisperse (black—continuous line) and polydisperse (red—dotted line) scattering intensity (**b**), (**d**), (**f**) at iteration $m = 4$ (see Refs. [22, 23]). Upper row: model MFI ($D_0 \simeq 1.24$). Middle row: model MFII ($D_0 \simeq 1.35$). Lower row: model MFIII ($D_0 \simeq 1.44$). Vertical lines in scattering intensity figures indicate the edges of the fractal regions. Here, D_0 is the box-counting fractal dimension, and a is the overall size of the fractal

noticed, which up to a certain extent, can be described as having a periodicity related to β_{s1}.

For models MFII and MFIII, and which correspond to cases when the scaling factors are relatively closer to each other, the gaps between the main distance-groups looks to be overlapped and are less pronounced, although some periodicity can be still observable, as shown in Fig. 4.24c and e. Here also the gaps within a single group are hardly observable. Occurrence of this overlapping is due to the close values of the scaling factors, since the position of the gaps corresponding to a scaling factor start to "interfere" with the positions of those corresponding to the other one. Due to this effect a more homogeneous structure is obtained. This is clearly reflected by the almost constant line in the dimension spectra of model MFII (the green curve from Fig. 4.23). Note that for 3D structures, the number of pair of distances for a given value of r/a is much higher than in 2d case, and thus the minima and maxima are more pronounced. Thus, the missing of a more clear periodicity in pddf is also an effect of the Euclidean dimensionality of the structure.

For the presented Vicsek-like multifractal models, the positions of the four disks with scaling factor β_{s1} can be described by:

$$G_1(\boldsymbol{q}) = \cos\left(q_x l_0 \left(1 - \beta_{s1}/2\right)\right) \cos\left(q_y l_0 \left(1 - \beta_{s1}/2\right)\right), \qquad (4.77)$$

while the position of the single disk with scaling factor β_{s2} is given by:

$$G_2(\boldsymbol{q}) = 1. \qquad (4.78)$$

Since total number of disks at mth iteration is given by Eq. (4.75), with $k_1 = 4$ and $k_2 = 1$, then the total surface area is $\left(k_1 \beta_{s1}^2 + k_2 \beta_{s2}^2\right)^m$. Explicitly, at $m = 1$ the fractal consists of k_1 disks of radius $r_1 = \beta_{s1} r_0$ and one disk of radius $r_2 = \beta_{s2} r_0$, with $r_0 = l_0/2$. Therefore, the corresponding form factor can be written as [22]:

$$F_1(\boldsymbol{q}) = \frac{k_1 \beta_{s1}^2 G_1(\boldsymbol{q}) F_0(\beta_{s1} \boldsymbol{q}) + k_2 \beta_{s2}^2 G_2(\boldsymbol{q}) F_0(\beta_{s2}^2 \boldsymbol{q})}{k_1 \beta_{s1}^2 + k_2 \beta_{s2}^2}, \qquad (4.79)$$

where $F_0(q)$ is the form factor of the initiator given by Eq. (4.13).

For iteration number $m = 2$, one repeats the same procedure for each disk, and obtains k_2 disks of radius $\beta_{s2}^2 r_0$, k_1 disks of radii $\beta_{s1} \beta_{s2} r_0$, k_1^2 disks of radii $\beta_{s2}^1 r_0$, k_1 and so on. Thus, at an arbitrarily m, the fractal form factor can be written in terms of a recurrence relation, as:

$$F_m(\boldsymbol{q}) = \frac{k_1 \beta_{s1}^2 G_1(\boldsymbol{q}) F_{m-1}(\beta_{s1} \boldsymbol{q}) + k_2 \beta_{s2}^2 G_2(\boldsymbol{q}) F_{m-1}(\beta_{s2}^2 \boldsymbol{q})}{k_1 \beta_{s1}^2 + k_2 \beta_{s2}^2}, \qquad (4.80)$$

Figure 4.24b, d and e shows the corresponding monodisperse (black curves) and polydisperse (red curves) form factors for the models MFI, MFII and respectively

MFIII at $m = 4$. In calculating the influence of polydispersity we have used a log-normal distribution function with the relative variance $\sigma_r = 0.2$ (see Ref. [4]).

The results show the presence of four main regions in the scattering curve for each model. For $ql_0 \simeq \pi$ we obtain a Guinier region, for $\pi \lesssim ql_0 \lesssim 2\pi/\beta_{s2}^m$ we have a mass fractal region with $I(q) \propto q^{-D_0}$, for $2\pi/\beta_{s2}^m \lesssim ql_0 2\pi/\beta_{s1}^m$ we have a surface-like fractal region with $I(q) \propto q^{-(4-D_0)}$, and for $2\pi/\beta_{s1}^m \lesssim ql_0$ the Porod region with $I(q) \propto q^{-3}$. Here, where D_0 is the box counting dimension of the multifractal, and whose value coincides with that obtained from the dimension spectra (see Fig. 4.23).

These observations show that the edges of any of these regions depends on numerical values of both scaling factors. The figures shown illustrate three important cases. First, for the model MFI, since $\beta_{s2} = 0.76$ is much larger than $\beta_{s1} = 0.12$, the upper edge of the mass fractal region is very close to its lower edge, so that no oscillations can be observed in this region (Fig. 4.24b). Thus, the length of the surface fractal region is relatively much larger, and a log-periodicity of the scattering curve with the period $\log(1/\beta_{s1})$ can be observed. Thus, behaviour of the SAS curve over the whole investigated range resembles that of scattering from a surface fractal. When the scaling factors are relatively closer to each other (model MFII), the corresponding scattering curve is shown in Fig. 4.24d and it is characterized by the presence of both mass and surface fractal regions of comparable lengths. This can describe typical experimental data from hierarchical systems consisting of of a succession from mass to surface power-law decay. Finally, when the scaling factors are very close to each other (model MFIII), a very narrow surface fractal region arise together with a much longer mass fractal region, as shown in Fig. 4.24f. This is to be expected, since when $\beta_{s1} = \beta_{s2}$ the multifractal is reduced to a single scale fractal, where only a mass-fractal region can be seen.

The results shown in Figs. 4.23 and 4.24 show that for models MFII and MFIII the relationships between the log-periodicity and the scaling factors is not so clearly visible as for model MI. This effect arise due to the superposition of maxima and minima, occurring from the 'mixing' of various structures of comparable sizes and distances between them. Therefore, a combined structural investigation involving SAXS/SANS experimental data as well as an image analysis of multifractals can be used to determine structural properties from both reciprocal and real space. While in the later case the phase is not lost and thus the structure can be directly obtained, in the former case the information is obtained from a macroscopic, statistically significant volume.

4.5.1 Case Studies

We present in this section two systems for which we calculate the dimension spectra and the scattering intensity. First, we simulate numerically the properties of 2D diffusion-limited aggregation (DLA) clusters, and then we investigate the surface of a membrane based silicone rubber (SR), stearic acid (SA) solution, and catalyst (C).

The first system provides results relevant to a large class of objects, while the second one illustrates the methodology presented in previous sections.

4.5.1.1   2D Diffusion-Limited Aggregation

In a DLA process the particles undergo a random walk and they cluster together to form aggregates. The theory of DLA has been developed in Ref. [24] to model systems where diffusion is the primary mean of transport, and thus is very useful to model growth processes, such as mineral deposits or carbon deposition on the walls of a cylinder of an engine. The clusters formed in a DLA process have a fractal structure, with fractal dimension about 1.71 in 2D.

Figure 4.25a shows a 2D DLA cluster consisting of 11266 particles [25]. The fractal nature of this structure is revealed through the existence of branched structures at various scales of magnification. The corresponding dimension spectrum is presented in Fig. 4.25b which shows a slightly decreasing behaviour of D_s, with $D_0 \simeq 1.68$ (horizontal thick-green line). This is very close to the theoretically estimated value, and which is $D_0 \simeq 1.71$ (horizontal blue—dotted line). The $p(r)$ function for the DLA is presented in Fig. 4.25c (black) together with that of a disk of the same size (red), in order to reflect the degree of asymmetry of the DLA. The number of distances used is about 63×10^6. The results show that both $p(r) \to 0$ at $r \simeq 471$ nm, and which gives the maximum distance (D_{max}) inside these structures (see also Ref. [25]). Note that, without loosing from generality, here we choose nm as units of length in order to covert the SAS experimental range. The pair distribution functions have a Gaussian-like shape with a full width at half maximum smaller for DLA, which arise due to a narrower range of values of distances between arbitrarily points in DLA. The maximum of DLA pair distribution function is situated at higher values, as compared to those of the disk, and indicates a relatively higher number of most common distances inside DLA. Figure 4.25d shows the corresponding scattering curves for DLA (black) and disk (red). The main feature is the presence of the Guinier region at $q \lesssim 2\pi/D_{max} \simeq 0.12$ nm^{-1}, followed by a fractal region with scattering exponent 1.71 for DLA, and exponent 2 for disks, as expected. These results show that in the proposed approach, the obtained structural parameters are very close to the theoretical ones, and thus it can be applied also to physical systems (see next section).

4.5.1.2   Silicone Rubber-Based Membranes

Fabrication of this type of membranes has been described in details in Ref. [26]. All the components are commercially available, and here we summarize only the main steps followed in their fabrication. Basically, they involve first preparation of a SA and catalyst-based solution by mixing SA (from Merck) with silicone oil (from Merck) with a ratio of volume fractions 4 : 96 at 350 K for about 10 min. When the temperature reaches 300 K, a CATA 6H type (from Bluestar-Silicone) catalyst is added to SA solution. Then, a homogeneous solution is prepared consisting of

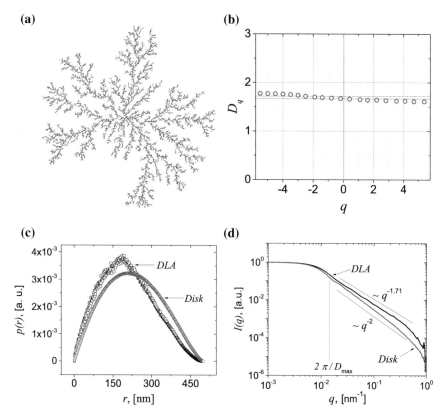

Fig. 4.25 a 2D DLA with 11266 particles. **b** DLA dimension spectrum. At $s = 0$ (vertical continuous thick-green line) we have $D_0 \simeq 1.67$ (horizontal continuous thick-green line). Theoretical values $D_0 = 1.71$ is marked with a horizontal dotted-blue line. **c** Pair distribution function of DLA (black) and of a disk of the same diameter D_{max} (red), for comparison. **d** Scattering intensity of DLA (black) and of a disk of the same diameter D_{max}. Vertical dotted line indicates the end of the Guinier region, and beginning of the fractal one. (see Ref. [25])

5×10^{-6} m^3 of SR, RTV-3325 type (from Bluestar-Silicone), mixed with 1×10^{-6} m^3 of SA and catalyst-solution at 336 K. At the end of the polymerization process, which lasts about 5 min., a porous polymeric membrane is obtained.

An image obtained with an optical microscope is presented in Fig. 4.26a, and shows the formation of open pores with various sizes on the surface of the membrane, either dead-end or through, and generally in the sub-micrometer range or higher. The physical mechanisms for their formation have been explained in Ref. [26], and is beyond the topic of our discussion. Figure 4.26b shows a binarized version of the optical microscopy image, with the purpose to underline more clearly the distribution of pores within the membrane's surface. For this purpose, a suitable threshold has been used based on a visual inspection of the image histogram. The image shows a random-like distribution of pores of various shapes and with a non-uniform density

(a)   **(b)**

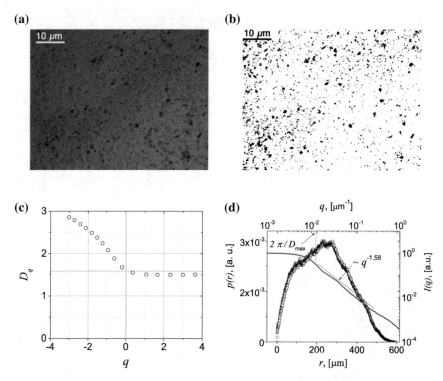

Fig. 4.26 a Optical microscopy (greyscale image) of a porous silicone rubber membrane. **b** The binarized version of the same membrane. **c** The corresponding dimension spectrum. The box-counting dimension is at $D_0 \simeq 1.58$. **d** Pair distribution function $p(r)$ versus r (black—discrete points) and the corresponding scattering intensity $I(q)$ versus q (red—continuous line). The vertical dotted line indicates the end of the Guinier region and the beginning of the fractal one, where $D_{max} \simeq 600 \, \mu m$

distribution. The average size of the pores is about $1 \, \mu m$. The dimension spectra is shown in Fig. 4.26c, which shows that in a reliable range of interest, i.e. when $-1 \lesssim q \lesssim 3$, D_s has a decreasing behavior for $-1 \lesssim s \lesssim 0$, followed by a quasi-constant one, at $0 \lesssim s \lesssim 3$. In the former s-range, the decay is quite pronounced, showing a relatively high degree of heterogeneity, while the asymptotic behavior reached in the later s-range reveal a mono-fractal like behaviour. At $s \simeq 0$ one obtains the box-counting dimension $D_0 \simeq 1.58$.

By considering the pores as the scattering objects, we calculate in Fig. 4.26d the pair distribution function ($p(r)$ vs. r; black-discrete points), and by a numerical Fourier transform, the corresponding scattering intensity ($I(q)$ vs. q; red-continuous line). The $p(r)$ function reveals that a maximum size of $D_{max} \simeq 600 \, \mu m$ with the highest number of distances at $\approx 250 \, \mu m$. The knees appearing at 90, 350 and 420 μm reflect the non-uniformity in density distribution of pores. Finally, the scattering curve also reveals the size of the system through the end of Guinier region at $q \simeq 2\pi/D_{max} \simeq 0.01$ μm^{-1}. In the fractal region, the exponent of the power-law decay coincide with the box-counting dimension obtained from dimension spectra. Note that although the size of the system is quite large for a typical SAS experiment, it clearly illustrates the potential of a combined approach between multifractal analysis and SAS experiments.

References

1. Schmidt, P.W.: J. Appl. Cryst. **15**(5), 567 (1982)
2. Schmidt, P.W., Dacai, X.: Phys. Rev. A **33**(1), 560 (1986)
3. Teixeira, J.: J. Appl. Cryst. **21**(6), 781 (1988)
4. Cherny, A.Y., Anitas, E.M., Osipov, V.A., Kuklin, A.I.: Phys. Rev. E **84**(3), 036203 (2011)
5. Cherny, A.Y., Anitas, E.M., Kuklin, A.I., Balasoiu, M., Osipov, V.A.: J. Appl. Cryst. **43**(4), 790 (2010)
6. Martin, J.E., Hurd, A.J.: J. Appl. Cryst. **20**(2), 61 (1987)
7. Schmidt, P.W.: J. Appl. Cryst. **24**(5), 414 (1991)
8. Cherny, A.Y., Anitas, E.M., Osipov, V.A., Kuklin, A.I.: Phys. Chem. Chem. Phys. **19**(3), 2261 (2017)
9. Li, C., Zhang, X., Li, N., Wang, Y., Yang, J., Gu, G., Zhang, Y., Hou, S., Peng, L., Wu, K., Nieckarz, D., Szabelski, P., Tang, H., Wang, Y.: J. Amer. Chem. Soc. **139**(39), 13749 (2017)
10. Melnichenko, Y.B.: Small-Angle Scattering from Confined and Interfacial Fluids. Springer International Publishing, Cham (2016)
11. Pfeifer, P., Ehrburger-Dolle, F., Rieker, T.P., González, M.T., Hoffman, W.P., Molina-Sabio, M., Rodríguez-Reinoso, F., Schmidt, P.W., Voss, D.J.: Phys. Rev. Lett. (2002)
12. Balasoiu, M., Anitas, E.M., Bica, I., Osipov, V.A., Orelovich, O.L., Savu, D., Savu, S., Kuklin, A.I.: Opt. Adv. Mater. - Rapid. Commun. **2**, 730 (2008)
13. Cherny, A.Y., Anitas, E.M., Osipov, V.A., Kuklin, A.I.: IUCr. J. Appl. Cryst. **50**(3), 919 (2017)
14. Bellido, E.P., Bernasconi, G.D., Rossouw, D., Butet, J., Martin, O.J.F., Botton, G.A.: ACS Nano **11**(11), 11240 (2017)
15. Zhao, J., Shi, D., Lian, J.: Carbon **47**(10), 2329 (2009)
16. Headen, T.F., Boek, E.S., Stellbrink, J., Scheven, U.M.: Langmuir **25**(1), 422 (2009)
17. Golosova, A.A., Adelsberger, J., Sepe, A., Niedermeier, M.A., Lindner, P., Funari, S.S., Jordan, R., Papadakis, C.M.: J. Phys. Chem. C **116**(29), 15765 (2012)
18. Anitas, E.M.: Eur. Phys. J. B **87**, 139 (2014)
19. Anitas, E.M.: Rom. J. Phys. **60**, 647 (2015)
20. Anitas, E.M., Slyamov, A., Todoran, R., Szakacs, Z.: Nanoscale Res. Lett. **12**(1), 389 (2017)
21. Vicsek, T.: Fractal Growth Phenomena, 2nd edn. World Scientific, Singapore (1992)
22. Cherny, A.Y., Anitas, E.M., Osipov, V.A., Kuklin, A.I.: Phys. Chem. Chem. Phys. (2019). https://doi.org/10.1039/C9CP00783K
23. Anitas, E.M., Marcelli, G., Szakacs, Z., Todoran, R., Todoran, D.: Symmetry **11**(6), 806 (2019)
24. Witten, T.A., Sander, L.M.: Phys. Rev. Lett. **47**, 1400 (1981)
25. Anitas, E.M.: Nanomaterials **9**, 648 (2019)
26. Bica, I., Anitas, E.M., Balasoiu, M., Iordaconiu, L., Bunoiu, M., Averis, L.M.E.: Rom. J. Phys. **61**, 464 (2016)

Chapter 5
Conclusions and Outlook

Recent progress in nanotechnology and materials science has paved the way for fabrication of deterministic fractal and multifractal structures at nano- and microscales. Synthesis of various materials, such as nanocomposites, based on this type of structures leads to end products with highly improved physical (mechanical, thermal, optical, electric) characteristics. However, the increasing needs of modern society require advanced materials with predefined properties and functions, and thus a fine control of their properties, when subjected to external factors such as magnetic fields, temperature gradient or mechanical tensions, is of prime importance. Thus, at a fundamental level, one of the basic challenges is to understand the correlations between the physical and structural properties of these fractal-based materials. In this book we focus on determination of structural properties using SAS technique, and present and discuss the latest theoretical developments useful for this purpose.

In the following we review the main features which makes SAS one of the most suitable experimental technique in structural investigations of nanoscale materials we summarize the main type of structures for which the existing theoretical models can be applied, and list the main structural information which can be extracted for each type of fractals. We also underlines the relationship between a SAS curve and dimension spectra, and identify key advantages, as well as some limitations of the presented approach.

Besides the standard advantages offered by SAS over other structural techniques, such as providing statistically significant quantities averaged over a macroscopic volume, eliminating the requirement for a priori sample preparation, allowing for deuteration (in the case of SANS) or differentiating between mass and surface fractals (through the values of the scattering exponents in the fractal region), one can also use SAS to distinguish between the following basic classes:

- random and deterministic single scale fractals (Fig. 5.1a),
- regular and fat fractals (Fig. 5.1b),
- regular and multifractals (Fig. 5.1c),
- fractals where the size of the basic units is of the same order as the distances between them, and where the size is much smaller (Fig. 5.1d).

© The Author(s), under exclusive license to Springer Nature Switzerland AG 2019   113
E. M. Anitas, *Small-Angle Scattering (Neutrons, X-Rays, Light)*
from Complex Systems, SpringerBriefs in Physics,
https://doi.org/10.1007/978-3-030-26612-7_5

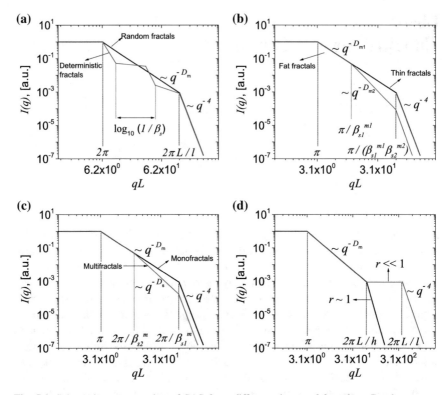

Fig. 5.1 Schematic representation of SAS from different classes of fractals. **a** Random versus deterministic. **b** Thin versus fat. where $m1$ and $m2$ are the number of fractal iteration in each structural level (see text for details). **c** Single versus two scale fractals with scaling factors $\beta_{s1} < \beta_{s2}$, at mth iteration. **d** $r \ll 1$ versus $r \simeq 1$ (see text for details). Here, h is the characteristic minimal distance between scattering units

In the first case, the differentiation is made through the type of the power-law decay in the fractal region, i.e. it is a simple power-law decay for random fractals, and a superposition of maxima and minima of increasing complexity on a power-law decay for deterministic ones (Fig. 5.1a). This type of superposition is known in the literature also as a generalized power-law decay. The same figure illustrates the main structural parameters which can be obtain from both classes. Generally, for random fractals one can obtain the fractal dimension D_m from the exponent of the power-law decay in the fractal region, the overall size L of the fractal from the end of Guinier region, and the size l of basic units from the end of fractal region. However, for deterministic fractals, in addition to D, L and l we can obtain the scaling factor β_s from the period of the curve in the fractal region, the number m of fractal iterations, which is equal to the number of periods, and the number of basic units N_m composing the fractal. In the limit of strong size polydispersity, the scattering curve of a deterministic fractal can be sufficiently smeared out such that it becomes indistinguishable from the scattering curve of a random fractal.

In the second case, while for regular (thin) fractals we have a single power-law decay (either simple or generalized), for fat fractals we have a succession of power-law decays with decreasing values of the scattering exponents. Thus, one can obtains also the fractal dimensions D_{m1} and D_{m2} of each structural level, i.e. from the regions with constant scaling factors β_{s1} and respectively β_{s2} (Fig. 5.1b). The length of each power-law decay can be regulated through the number of iterations for which the scaling factor is constant. Note that this is a particular case of a more general case, where a fractal consists from different structural levels. In this case, the exponents in SAS intensity not necessarily need to have decreasing values.

In the third case, for multifractals with two-scaling factors the scattering curve is generally also characterized by a succession of power-law decays but they correspond to a succession of mass-to-surface fractals, and not mass-to-mass as in the previous case (Fig. 5.1c). For two scale fractals, both scaling factors β_{s1} and β_{s1} can be obtained from the periodicity of the SAS curve in the corresponding fractal region, as for the regular single scale fractals. The length of these fractal regions is set by the relative values of the two scaling factors.

Lastly, for fractals in which the size of the basic units is comparable with the distance between them is reflected in scattering intensity through the presence of a small knee between the two power-law decays. However, when their size is relatively much smaller, the knee is replaced by a Guinier region, whose length is related to the ratio r of the size to their distances (Fig. 5.1d).

Note that due to instrumental limitations, experimental SAS data may contain only parts of the scattering curve, generally either Guinier and fractal regions, or fractal and Porod regions. Sometimes it may happens that only the fractal region is present. Thus, quite often, in order to obtain all the regions, several experiments are usually needed, such as a combination of SAS and USAS, followed by data merging.

The structural information obtained from SAS measurements are usually complemented from other methods such as scanning/transmission electron microscopy or computed tomography. In particular, when images of the specimen are available, a multifractal analysis can be performed, and the obtained results can be correlated with those from SAS spectra. For example, the box-counting dimension coincide with the scattering exponent of SAS intensity. However, an open question is whether the other dimensions from the dimension spectra can be recovered from scattering experiments. Although theoretical developments are underway to address this question, from a practical point of view several limitations still shall be overcome: the sample, the method and the instrumentation.

Concerning the samples, probably the biggest drawback results from their limited availability. Although in the the last years, several types of materials have been used to synthesize deterministic structures such as dicarbonitrile/benzene/bis-terpyridine/terphenyl molecules, single/poly-crystalline silicone or Fe atoms, they can not fulfill the increasing technological and industrial demands for synthesis of materials with a particular chemical composition or matrix-filler combination, mainly due to their high production costs. Moreover, since SAS measures a macroscopic volume, up to about 0.2 cm^3, a good signal requires about 10^6 deterministic fractals in each investigated volume. Therefore large scale synthesis of such structures is still a challenge, and their production is limited to research laboratories.

Although the deterministic models and the suggested receipts helps in extracting additional information as compared with classical random fractal models, there is still no obvious way which would allow to distinguish between fractals with the same fractal dimension, since at a given value of fractal dimension may correspond a large number of structures. This limitation is related to the loss of phase information, which is an inherent process in scattering experiments. Several approaches to overcome this have been suggested such as through correlating the SAS spectrum with the lacunarity or by employing a numerical procedure to recover the phase in coherent SAXS, based on ptychographic methods. In addition, depending on the type of the fractal, individual classes rise their own particular questions. How to extract all the scaling factors in SAS from multifractals and what is a general expression of the corresponding form factor for arbitrarily iteration number? Experimentally, depending on the chemical structure of the sample, one can employ the variation contrast methods in SANS to emphasize or conceal a given region in the sample. Each region with a given value of the scaling factor may be matched to correspond to a given scattering length, and thus by contrast variation, one can in principle determine the structural properties including the scaling factors. How to relate the properties of scattering curve of self-affine fractal with the structure in the real space, is another interesting open question. Therefore, theoretically such problems shall be investigated in depth.

Last, but not least, finite instrumental resolution leads to smearing the experimentally curves up to a certain degree. This effects arise arise due to the wavelength spread, finite collimation or detector resolution. For SAS from deterministic fractals this is a serious issue since this leads to a partial or total disappearance of the oscillations in the fractal region. In the later case we face with the impossibility to extract several parameters such as the scaling factors and the fractal iteration number, which otherwise are available in the existing deterministic models. Therefore, additional progress in SAS instrumentation is elimination to a higher degree the above mentioned drawbacks, without significantly extending the measurement time or sacrificing the complex morphology of the investigated material.

Printed in the United States
By Bookmasters